T0317522

INTRODUCTION TO STATISTICAL ANALYSIS OF LABORATORY DATA

INTRODUCTION TO STATISTICAL ANALYSIS OF LABORATORY DATA

ALFRED A. BARTOLUCCI

University of Alabama at Birmingham
Birmingham, Alabama, USA

KARAN P. SINGH

University of Alabama at Birmingham
Birmingham, Alabama, USA

SEJONG BAE

University of Alabama at Birmingham
Birmingham, Alabama, USA

Copyright © 2016 by John Wiley & Sons, Inc. All rights reserved

Published by John Wiley & Sons, Inc., Hoboken, New Jersey
Published simultaneously in Canada

No part of this publication may be reproduced, stored in a retrieval system, or transmitted in any form or
by any means, electronic, mechanical, photocopying, recording, scanning, or otherwise, except as
permitted under Section 107 or 108 of the 1976 United States Copyright Act, without either the prior
written permission of the Publisher, or authorization through payment of the appropriate per-copy fee to
the Copyright Clearance Center, Inc., 222 Rosewood Drive, Danvers, MA 01923, (978) 750-8400, fax
(978) 750-4470, or on the web at www.copyright.com. Requests to the Publisher for permission should
be addressed to the Permissions Department, John Wiley & Sons, Inc., 111 River Street, Hoboken, NJ
07030, (201) 748-6011, fax (201) 748-6008, or online at http://www.wiley.com/go/permission.

Limit of Liability/Disclaimer of Warranty: While the publisher and author have used their best efforts in
preparing this book, they make no representations or warranties with respect to the accuracy or
completeness of the contents of this book and specifically disclaim any implied warranties of
merchantability or fitness for a particular purpose. No warranty may be created or extended by sales
representatives or written sales materials. The advice and strategies contained herein may not be suitable
for your situation. You should consult with a professional where appropriate. Neither the publisher nor
authors shall be liable for any loss of profit or any other commercial damages, including but not limited
to special, incidental, consequential, or other damages.

For general information on our other products and services or for technical support, please contact our
Customer Care Department within the United States at (800) 762-2974, outside the United States at
(317) 572-3993 or fax (317) 572-4002.

Wiley also publishes its books in a variety of electronic formats. Some content that appears in print may
not be available in electronic formats. For more information about Wiley products, visit our web site at
www.wiley.com.

Library of Congress Cataloging-in-Publication Data:

Bartolucci, Alfred A., author.
  Introduction to statistical analysis of laboratory data / Alfred A. Bartolucci, Karan P. Singh, Sejong Bae.
    pages cm
  Includes bibliographical references and index.
  ISBN 978-1-118-73686-9 (cloth)
1. Diagnosis, Laboratory–Statistical methods. 2. Statistics. I. Singh, Karan P., author. II. Bae, Sejong,
author. III. Title.
  RB38.3.B37 2016
  616.07′50151–dc23

                                                                                                2015025700

Typeset in 10/12pt TimesLTStd by SPi Global, Chennai, India

Printed in the United States of America

10 9 8 7 6 5 4 3 2 1

1   2016

To Lieve and Frank

CONTENTS

PREFACE

INTENDED AUDIENCE

The advantage of this book is that it provides a comprehensive knowledge of the analytical tools for problem solving related to laboratory data analysis and quality control. The content of the book is motivated by the topics that a laboratory statistics course audience and others have requested over the years since 2003. As a result, the book could also be used as a textbook in short courses on quantitative aspects of laboratory experimentation and a reference guide to statistical techniques in the laboratory and processing of pharmaceuticals. Output throughout the book is presented in familiar software format such as EXCEL and JMP (SAS Institute, Cary, NC).

The audience for this book could be laboratory scientists and directors, process chemists, medicinal chemists, analytical chemists, quality control scientists, quality assurance scientists, CMC regulatory affairs staff and managers, government regulators, microbiologists, drug safety scientists, pharmacists, pharmacokineticists, pharmacologists, research and development technicians, safety specialists, medical writers, clinical research directors and personnel, serologists, and stability coordinators. The book would also be suitable for graduate students in biology, chemistry, physical pharmacy, pharmaceutics, environmental health sciences and engineering, and biopharmaceutics. These individuals usually have an advanced degree in chemistry, pharmaceutics, and formulation science and hold job titles such as scientist, senior scientist, principal scientist, director, senior director, and vice president. The above partial list of titles is from the full list of attendees that have participated in the 2-day course titled "Introductory Statistics for Laboratory Data Analysis" given through the Center for Professional Innovation and Education.

PROSPECTUS

There is an unmet need to have the necessary statistical tools in a comprehensive package with a focus on laboratory experimentation. The study of the statistical handling of laboratory data from the design, analysis, and graphical perspective is essential for understanding pharmaceutical research and development of results involving practical quantitative interpretation and communication of the experimental process. A basic understanding of statistical concepts is pertinent to those involved in the utilization of the results of quantitation from laboratory experimentation and how these relate to assuring the quality of drug products and decisions about bioavailability, processing, dosing and stability, and biomarker development. A fundamental knowledge of these concepts is critical as well for design, formulation, and manufacturing.

This book presents a detailed discussion of important basic statistical concepts and methods of data presentation and analysis in aspects of biological experimentation requiring a fundamental knowledge of probability and the foundations of statistical inference, including basic statistical terminology such as simple statistics (e.g., means, standard deviations, medians) and transformations needed to effectively communicate and understand one's data results. Statistical tests (one-sided, two-sided, nonparametric) are presented as required to initiate a research investigation (i.e., research questions in statistical terms). Topics include concepts of accuracy and precision in measurement analysis to ensure appropriate conclusions in experimental results including between- and within-laboratory variation. Further topics include statistical techniques to compare experimental approaches with respect to specificity, sensitivity, linearity, and validation and outlier analysis. Advanced topics of the book go beyond the basics and cover more complex issues in laboratory investigations with examples, including association studies such as correlation and regression analysis with laboratory applications, including dose response and nonlinear dose–response considerations. Model fit and parallelism are presented. To account for controllable/uncontrollable laboratory conditions, the analysis of robustness and ruggedness as well as suitability, including multivariate influences on response, are introduced. Method comparison using more accurate alternatives to correlation and regression analysis and pairwise comparisons including the Mandel sensitivity are pursued. Outliers, limit of detection and limit of quantitation and data handling of censored results (results below or above the limit of detection) with imputation methodology are discussed. Statistical quality control for process stability and capability is discussed and evaluated. Where relevant, the procedures provided follow the CLSI (Clinical and Laboratory Standards Institute) guidelines for data handling and presentation.

The significance of this book includes the following:

- A comprehensive package of statistical tools (simple, cross-sectional, and longitudinal) required in laboratory experimentation
- A solid introduction to the terminology used in many applications such as the interpretation of assay design and validation as well as "fit-for-purpose" procedures
- A rigorous review of statistical quality control procedures in laboratory methodologies and influences on capabilities
- A thorough presentation of methodologies used in the areas such as method comparison procedures, limit and bias detection, outlier analysis, and detecting sources of variation.

ACKNOWLEDGMENTS

The authors would like to thank Ms. Laura Gallitz for her thorough review of the manuscript and excellent suggestions and edits that she provided throughout.

1

DESCRIPTIVE STATISTICS

1.1 MEASURES OF CENTRAL TENDENCY

One wishes to establish some basic understanding of statistical terms before we deal in detail with the laboratory applications. We want to be sure to understand the meaning of these concepts, since one often describes the data with which we are dealing in summary statistics. We discuss what is commonly known as measures of central tendency such as the mean, median, and mode plus other descriptive measures from data. We also want to understand the difference between samples and populations.

Data come from the samples we take from a population. To be specific, a population is a collection of data whose properties are analyzed. The population is the *complete* collection to be studied; it contains all possible data points of interest. A sample is a part of the population of interest, a subcollection selected from a population. For example, if one wanted to determine the preference of voters in the United States for a political candidate, then all registered voters in the United States would be the population. One would sample a subset, say, 5000, from that population and then determine from the sample the preference for that candidate, perhaps noting the percent of the sample that prefer that candidate over another. It would be impossible logistically and costwise in statistics to canvass the entire population, so we take what we believe to be a representative sample from the population. If the sampling is done appropriately, then we can generalize our results to the whole population. Thus, in statistics, we deal with the sample that we collect and make our decisions. Again, if

Introduction to Statistical Analysis of Laboratory Data, First Edition.
Alfred A. Bartolucci, Karan P. Singh, and Sejong Bae.
© 2016 John Wiley & Sons, Inc. Published 2016 by John Wiley & Sons, Inc.

we want to test a certain vegetable or fruit for food allergens or contaminants, we take a batch from the whole collection, send it to the laboratory and it is, thus, subjected to chemical testing for the presence or degree of the allergen or contaminants. There are certain safeguards taken when one samples. For example, we want the sample to appropriately represent the whole population. Factors relevant in considering the representativeness of a sample include the homogeneity of the food and the relative sizes of the samples to be taken, among other considerations. Therefore, keep in mind that when we do statistics, we always deal with the sample in the expectation that what we conclude generalizes to the whole population.

Now let's talk about what we mean when we say we have a distribution of the data. The following is a sample of size 16 of white blood cell (WBC) counts ×1000 from a diseased sample of laboratory animals:

$$5.13, 5.4, 5.4, 5.7, 5.7, 5.7, 6.0, \underline{6.0, 6.0}, 6.0, 6.13, 6.13, 6.13, 6.4, 6.4, 6.8.$$

Note that this data is purposely presented in ascending order. That may not necessarily be the order in which the data was collected. However, in order to get an idea of the range of the observations and have it presented in some meaningful way, it is presented as such. When we rank the data from the smallest to the largest, we call this a distribution.

One can see the distribution of the WBC counts by examining Figure 1.1. We'll use this figure as well as the data points presented to demonstrate some of the statistics that will be commonplace throughout the text. The height of the bars represents the frequency of counts for each of the values 5.13–6.8, and the actual counts are placed on top of the bars. Let us note some properties of this distribution. The mean is easy. It is obviously the average of the counts from 5.13 to 6.8 or $(5.13 + 5.4 + \cdots + 6.8)/16 = 5.939$. Algebraically, if we denote the elements of a sample of size

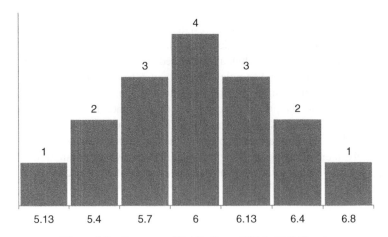

Figure 1.1   Frequency Distribution of White Cell Counts

n as X_1, X_2, ... , X_n, then the sample mean in statistical notation is equal to

$$\overline{X} = \frac{(X_1 + X_2 + \cdots + X_n)}{n}. \tag{1.1}$$

For example, in our aforementioned WBC data, $n = 16$, $X_1 = 5.13, X_2 = 5.4$, and so on, where $X_{16} = 6.8$.

Then the mean is noted as earlier, $(5.13 + 5.4 + \cdots + 6.8)/16 = 5.939$.

The median is the middle data point of the distribution when there is an odd number of values and the average of the two middle values when there is an even number of values in the distribution. We demonstrate it as follows.

Note our data is:

$$5.13, 5.4, 5.4, 5.7, 5.7, 5.7, 6.0, 6.0, \underline{6.0, 6.0}, 6.13, 6.13, 6.13, 6.4, 6.4, 6.8.$$

The number of data points is an even number, or 16. Thus, the two middle values are in positions 8 and 9 underlined above. So the median is the average of 6.0 and 6.0 or $(6.0 + 6.0)/2 = 12.0/2 = 6.0$ or median $= 6.0$.

Suppose we had a distribution of seven data points, which is an odd number, then the median is just the middle value or the value in position number 4. Note the following: $5.13, 5.4, 5.6, 5.7, 5.8, 5.8, 6.0$. Thus, the median value is 5.7. The median is also referred to as the 50th percentile. Approximately 50% of the values are above it and 50% of the values are below it. It is truly the middle value of the distribution.

The mode is the most frequently occurring value in the distribution. If we examine our full data set of 16 points, one will note that the value 6.0 occurs four times. Also see Figure 1.1. Thus, the mode is 6.0. One can have a distribution with more than one mode. For example, if the values of 5.4 and 6.0 were each counted four times, then this would be a bimodal distribution or a distribution with two modes.

We have just discussed what is referred to as measures of central tendency. It is easy to see that the measures of central tendency from this data (mean, median, and mode) are all in the center of the distribution, and all other values are centered around them. In cases where the mean = median = mode as in our example, the distribution is seen to be symmetric. Such is not always the case.

Figure 1.2 deals with data that is skewed and not symmetric. Note the mode to the left indicating a high frequency of low values. These are potassium values from a laboratory sample. This data is said to be skewed to the right or positively skewed. We'll revisit this concept of skewness in Chapter 2 and later chapters as well. There are 23 values (not listed here) ranging from 30 to 250. One usually computes the geometric mean (GM) of the data of this form. Sometimes, GM is preferred to the arithmetic mean (ARM) since it is less sensitive to outliers or extreme values. Sometimes, it is called a "spread preserving" statistic. The GM is always less than or equal to the ARM and is commonly used with data that may be skewed and not normal or not symmetric, such as much laboratory data is not symmetric.

Suppose we have n observations X_1, X_2, ... , X_n , then the GM is defined as

$$GM = \prod_{i=1,n} X_i^{1/n} = X_1^{1/n} X_2^{1/n} X_n^{1/n}, \tag{1.2}$$

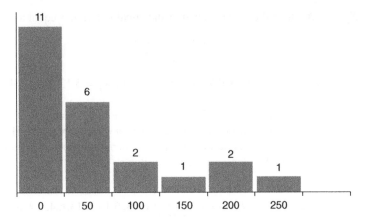

Figure 1.2   Frequency Distribution of Potassium Values

or equivalently

$$GM = \exp\left[\frac{(\log X_1 + \log X_2 + \cdots + \log X_n)}{n}\right]. \qquad (1.3)$$

In our potassium example $GM = 30^{1/23}35^{1/23}, \ldots, 250^{1/23} = 65.177$. Note that the $ARM = 75.217$.

1.2   MEASURES OF VARIATION

We've learned some important measures of statistics. The mean, median, and mode describe some sample characteristics. However, they don't tell the whole story. We want to know more characteristics of the data with which we are dealing. One such measure is the dispersion or the variance. This particular measure has several forms in laboratory science and is essential to determining something about the precision of an experiment. We will discuss several forms of variance and relate them to data accordingly.

The range is the difference between the maximum and minimum value of the distribution. Referring to the WBC data:

$$Range = maximum\ value\text{–}minimum\ value = 6.8 - 5.13 = 1.67.$$

Obviously, the range is easy to compute, but it only depends on the two most extreme values of the data. We want a value or measure of dispersion that utilizes all of the observations. Note the data in Table 1.1. For the sake of demonstration, we have three observations: 2, 4, and 9. These data are seen in the data column.

TABLE 1.1 Demonstration of Variance

Observation	Data	Deviation	(Deviation)^2
1	2	$(2-5)=-3$	$(2-5)^2=9$
2	4	$(4-5)=-1$	$(4-5)^2=1$
3	9	$(9-5)=4$	$(9-5)^2=16$
Sum	15	0	26
Average	5	0	$26/(3-1)=13$

Note their sum or total is 15. Their mean or average is 5. Note their deviation from the mean, $2-5=-3$, $4-5=-1$ and $9-5=4$. The sum of their deviations is 0. This property is true for any size data set, that is, the sum of the deviations will be close to 0. This doesn't make much sense as a measure of dispersion or we would have a perfect world of no variation or dispersion of the data. The last column denoted as (Deviation)^2 is the deviation column squared. And the sum of the squared deviations is 26.

The variance of a sample is the average squared deviation from the sample mean. Specifically, from the previous sample of three values, $[(2-5)^2 + (4-5)^2 + (9-5)^2]/(3-1) = [9 + 1 + 16]/2 = 26/2 = 13$. Thus, the variance is 13. Dividing by $(3-1)=2$ instead of 3 gives us an *unbiased* estimator of the variance because it tends to closely estimate the true population variance. Note that if our sample size were 100, then dividing by 99 or 100 would not make much of a difference in the value of the variance. The adjustment of dividing the sum of squares of the deviation by the sample size minus 1, $(n-1)$, can be thought of as a small sample size adjustment. It allows us not to underestimate the variance but to conservatively overestimate it.

Recall our WBC data:

$$5.13, 5.4, 5.4, 5.7, 5.7, 5.7, 6.0, 6.0, 6.0, 6.0, 6.13, 6.13, 6.13, 6.4, 6.4, 6.8.$$

The mean or average is: $5.939 = 5.94$.
So the variance is

$$\text{Var} = \frac{[(5.13 - 5.94)^2 + (5.4 - 5.94)^2 + \cdots + (6.8 - 5.94)^2]}{15} = 0.1798$$

Algebraically, one may note the variance formula in statistical notation for the data in Table 1.1, where the mean is $\overline{X} = 5$.
One defines the sample variance as S^2_{n-1} or

$$S^2_{n-1} = \frac{\sum (X_i - \overline{X})^2}{n-1} \tag{1.4}$$

So for the data in Table 1.1 we have

$$S^2_{n-1} = \frac{(X_1 - \overline{X})^2 + (X_2 - \overline{X})^2 + (X_3 - \overline{X})^2}{3 - 1}$$

$$= \frac{(2-5)^3 + (4-5)^2 + (9-5)^2}{2} = \frac{(-3)^2 + (-1)^2 + (4)^2}{2}$$

$$= \frac{9 + 1 + 16}{2} = \frac{26}{2} = 13$$

The sample standard deviation (SD), S_{n-1}, is the square root of sample variance $= \sqrt{S^2_{n-1}}$, or in our case $\sqrt{13} = 3.606$.

$$\sqrt{S^2_{m-1}} = \sqrt{13} = 3.606 \tag{1.5}$$

The variance is a measure of variation. The square root of the variance, or SD, is a measure of variation in terms of the original scale.

Thus, referring back to the aforementioned WBC data, the SD of our WBC counts is the square root of the variance, that is, $\sqrt{0.1798} = 0.4241$.

Just as we discussed the GM earlier for data that may be possibly skewed, we also have a geometric standard deviation (GSD). One uses the log of the data as we did for the GM. The GSD is defined as

$$GSD = \exp\left[\sqrt{\frac{\sum_{i=1,n}\{\log X_i - \log GM\}^2}{n - 1}}\right]. \tag{1.6}$$

As an example, suppose we have $n = 10$ data points $100, 99, 100, 90, 90, 70, 89, 70, 64, 56$.

Then from (1.6), the GSD $= 1.233$. Unlike the GM, the GSD is not necessarily a close neighbor of the arithmetic SD, which in this case is 16.315.

Another measure of variation is the standard error of the mean (SE or SEM), which is the SD divided by the square root of the sample size or

$$SE = \frac{SD}{\sqrt{n}}. \tag{1.7}$$

For our aforementioned WBC data, we have $SE = 0.4241/\sqrt{16} = 0.4241/4 = 0.1060$.

The standard error (SE) of the mean is the variation one would expect in the sample means after repeated sampling from the same population. It is the SD of the sample means. Thus, the sample SD deals with the variability of your data while the SE of the mean deals with the variability of your sample mean.

Naturally, we have only one sample and one sample mean. Theoretically, the SE is the SD of many sample means after sampling repeatedly from the same population. It can be thought of as a SD of the sample means from replicated sampling or experimentation. Thus, a good approximation of the SE of the mean from one sample is the SD divided by the square root of the sample size as seen earlier. It is naturally smaller than the SD. This is because from repeated sampling from the population one would not expect the mean to vary much, certainly not as much as the sample data. Rosner (2010, Chapter 6, Estimation) and Daniel (2008, Chapter 6, Estimation) give an excellent demonstration and explanation of the SD and SE of the mean comparisons.

Another common measure of variation used in laboratory data exploration is the coefficient of variation (CV), sometimes referred to as the relative standard deviation (RSD). This is defined as the ratio of the SD to the mean expressed as a percent.

It is also called a measure of reliability – sometimes referred to as precision and is defined as

$$CV = \left(\frac{SD}{mean} \right) \times 100. \tag{1.8}$$

Our Sample CV of the WBC measurements is $CV = \left(\frac{0.4241}{5.94} \right) \times 100 = 7.14$.

The multiplication by 100 allows it to be referred to as the percent CV, %CV, or CV%.

The %CV normalizes the variability of the data set by calculating the SD as a percent of the mean. The %CV or CV helps one to compare the precision differences that may exist among assays and assay methods. We'll see an example of this in the following section. Clearly, an assay with $CV = 7.1\%$ is more precise than one with $CV = 10.3\%$.

1.3 LABORATORY EXAMPLE

The following example is based on the article by Steele et al. (2005) from the *Archives of Pathology and Laboratory Medicine*. The objective of the study was to determine the long-term within- and between-laboratory variation of cortisol, ferritin, thyroxine, free thyroxine, and Thyroid-Stimulating Hormone (TSH) measurements by using commonly available methods and to determine if these variations are within accepted medical standards, that is to say within the specified CV.

The design – Two vials of pooled frozen serum were mailed 6 months apart to laboratories participating in two separate College of American Pathologists' surveys. The data from those laboratories that analyzed an analyte in both surveys were used to determine for each method the total variance and the within- and between-laboratory variance components. For our purposes, we focus on the CV for one of the analytes, namely, the TSH. There were more than 10 analytic methods studied in this survey. The three methods we report here are as follows: A – Abbott AxSYM, B – Bayer

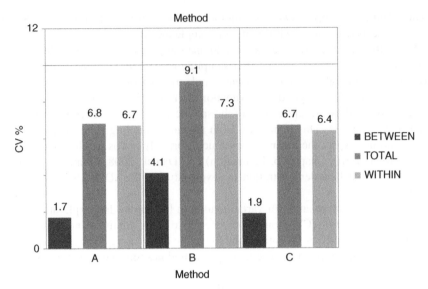

Figure 1.3 CV% for TSH. Reproduced in part from Steele et al. (2005) with permission from Archives of Pathology and Laboratory Medicine. Copyright 2005 College of American Pathologists

Advia Centaur, and C – Bayer Advia Centaur 3G. The study examined many end-points directed to measuring laboratory precision with a focus on total precision overall and within- and between-laboratory precision. The within-laboratory goals as per the %CV based on biological criteria were cortisol – 10.43%, ferritin – 6.40%, thyroxine – 3.00%, free thyroxine – 3.80%, and TSH – 10.00%. Figure 1.3 shows the graph for analytic methods A, B, and C, for TSH. The horizontal reference line across the top of the figure at 10% indicates that all of the bars for the total, within- and between-laboratory %CV met the criteria for the three methods shown here. Also, note in examining Figure 1.3 that the major source of variation was within-laboratory as opposed to the between- or among-laboratory variation or %CV.

When examining the full article, the authors point out that the number of methods that met within-laboratory imprecision goals based on biological criteria were 5 of 5 for cortisol; 5 of 7 for ferritin; 0 of 7 for thyroxine and free thyroxine; and 8 of 8 for TSH. Their overall conclusion was that for all analytes tested, the total within-laboratory component of variance was the major source of variation. In addition, note that there are several methods, such as thyroxine and free thyroxine that may not meet analytic goals in terms of their imprecision.

1.4 PUTTING IT ALL TOGETHER

Let's consider a small data set of potassium values and demonstrate summary statistics in one display. Table 1.2 gives the potassium values denoted by the X_i, where

TABLE 1.2 Potassium Values and Descriptive Statistics

i	X_i	$Y_i = \ln(X_i)$
1	3.2	1.163
2	4.2	1.435
3	3.8	1.335
4	2.3	0.833
5	1.5	0.405
6	15.5	2.741
7	8.5	2.140
8	7.9	2.067
9	3.1	1.131
10	4.4	1.482
Mean	$\overline{X} = 5.44$	$\overline{Y} = 1.47$
Variance	$S^2_X = 17.53$	$S^2_Y = 0.46$
Standard deviation	$SD_X = 4.19$	$SD_Y = 0.68$
Standard error of the mean	$SE_X = 1.32$	$SE_Y = 0.22$

$i = 1, 2, \ldots, 10$. The natural log of the values are seen in the third column denoted by $Y_i = \ln(X_i)$. The normal range of values for adult laboratory potassium (K) levels are from 3.5–5.2 milliequivalents per liter (mEq/L) or 3.5–5.2 millimoles per liter (mmol/L). Obviously, a number of the values are outside the range. The summary statistics are provided for both raw and transformed values, respectively. The Y values are actually from what we call a log-normal distribution, which we will discuss in the following chapter. Focusing on the untransformed potassium values of Table 1.2, Table 1.3 gives a complete set of summary statistics that one often encounters. We've discussed most of them and will explain the others. The minimum and maximum values are obvious, being the minimum and maximum potassium values from Table 1.2. The other two added values in Table 1.3 are 25th percentile (first quartile) and 75th percentile (third quartile). They are percentiles just like the median. Just as the median is the 50th percentile (second quartile) in which approximately 50% of the values may lie above it as well as below it, the 25th percentile is the value of 2.9, meaning that approximately 25% of the values in the distribution lie below it, which implies about 75% of the values in the distribution lie above the value 2.9. Thus, the 75th percentile is the value of 8.05, meaning that 75% of the values in the distribution are less than or equal to 8.05, implying that about 25% of the values lie above it. Note that the median is in the middle of the 25th and 75th percentile. These values between the 25th and 75th quartile are called the interquartile range (IQR). Note that approximately 50% of the data points are in the IQR.

Let's revisit the GM and GSD. From Table 1.2, we note that

$$GM = \exp\left[\frac{(\log X_1 + \log X_2 + \cdots + \log X_n)}{n} \right] = \exp[\overline{Y}] = \exp[1.47] = 4.349.$$

**TABLE 1.3   Descriptive Statistics of
10 Potassium (X) Values**

Mean	5.44
Standard deviation	4.19
Standard error mean	1.32
N	10
Minimum	1.5
25th percentile	2.9
Median	4.0
75th percentile	8.05
Maximum	15.5

Also, the relation between the arithmetic standard and GSD is such that ln (GSD) = arithmetic SD of the $Y_i;s$ in Table 1.2. Thus, ln(GSD) = 0.68 or GSD = exp(0.68) = 1.974.

1.5   SUMMARY

We have briefly summarized a number of basic descriptive statistics in this chapter such as the measures of central tendency and measures of variation. We also put them in the context of data that has a symmetric distribution as well as data that is not symmetrically distributed or may be skewed. It is important to note that these statistics just describe some property of the sample with which we are dealing in laboratory experimentation. Our goal in the use of these statistics is to describe what is expected to be true in the population from which the sample was drawn. In the next chapter, we discuss inferential statistics, which leads us to draw scientific conclusions from the data.

REFERENCES

Daniel WM. (2008). Biostatistics: A Foundation for Analysis in the Health Sciences, 9th ed., John Wiley & Sons, New York.

Rosner B. (2010). Fundamentals of Biostatistics, 7th ed., Cengage Learning.

Steele BW, Wang E, Palmer-Toy DE, Killeen AA, Elin RJ and Klee GG. (2005). Total long-term within-laboratory precision of cortisol, ferritin, thyroxine, free thyroxine, and Thyroid-Stimulating Hormone (TSH) assays based on a College of American Pathologists fresh frozen serum study: do available methods meet medical needs for precision? Archives of Pathology and Laboratory Medicine 129(3): 318–322.

2

DISTRIBUTIONS AND HYPOTHESIS TESTING IN FORMAL STATISTICAL LABORATORY PROCEDURES

2.1 INTRODUCTION

In this chapter, we review hypothesis testing and related terminology (type I and type II errors, *p*-value, etc.). Not all data follow the normal distribution and many laboratory measures are not symmetric or normal, and are in fact somewhat skewed. One should then be able to differentiate between the parametric approach (assuming that the data is normal or some other known distribution) and the nonparametric approach (assuming no known underlying distribution or assuming the data is not symmetric or normal) when performing inferential statistical tests. The *t*-distribution is discussed for two independent normal sample comparisons, and the analysis of variance (ANOVA) is discussed for multisample (≥ 3) comparisons. Nonparametric techniques (Wilcoxon and Kruskal–Wallis) are introduced as an alternative test in the case of skewed (not normal) data. This chapter also includes discussion of confidence intervals (CIs) and the relation of the CI to the hypothesis test. Of particular note is the relevance of the CI in the measure of precision.

Suppose we have 116 BUN (blood urea nitrogen) values from a Northeastern Laboratory. The distribution is shown in Figure 2.1. The descriptive summary of the data is found in Table 2.1. In this example, the distribution has the three central tendency values to the near equality: Median (28.5), Mean (28.4), and Mode (28.8). The distribution is nearly symmetric. Obviously, not all distributional shapes are symmetric.

Let's take a look at the shape of the distribution in Figure 2.2. Depending on the shape of the distribution, it can be referred to as positively (right) skewed, negatively

Introduction to Statistical Analysis of Laboratory Data, First Edition.
Alfred A. Bartolucci, Karan P. Singh, and Sejong Bae.
© 2016 John Wiley & Sons, Inc. Published 2016 by John Wiley & Sons, Inc.

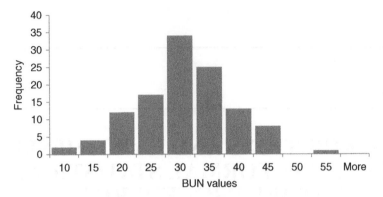

Figure 2.1   Distribution of 116 BUN Values – Northeast Lab

TABLE 2.1   Descriptive Statistics of 116 BUN Values – Northeast Lab

Mean	28.364
Standard deviation	7.889
Standard error	0.736
N	116
Mode	28.8
Minimum	7.304
25th percentile	24.36
Median	28.452
75th percentile	33.252
Maximum	51.464

(left) skewed, or symmetric distribution. The shape of the distribution can be determined by looking at the shape of data based on the central tendency measures. Right (positive) skewed data can be described as having a longer right tail or extreme values to the right as in Figure 2.3.

The descriptive summary of the right skewed data is found in Table 2.2. These are potassium values from the same laboratory. Note the relationship of the central tendency measures: Mode(35) < Median(92.50) < Mean(121.97). The mean value is sensitive to the extreme values and tends to pull toward the extreme value direction while the median and mode are less sensitive.

Left (negative) skewed data can be described as having a longer left tail or extreme values to the left as in the cytometry values shown in Figure 2.4. The descriptive summary of the left skewed data is found in Table 2.3. The relationship of the central tendency measures is Mean(249.1) < Median(271.5) < Mode(337). Again, the mean value is more sensitive to the extreme values and tends to pull toward the extreme value direction while the median and mode are less sensitive. Note that the direction of the skewness determines the position of the mean and mode, but the median is always in the center.

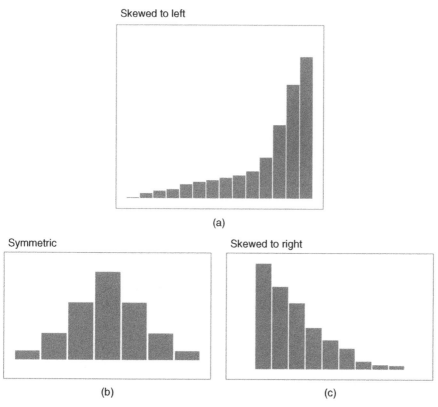

Figure 2.2 Shape of the Distribution: (a) Skewed to Left. (b) Symmetric. (c) Skewed to Right

Let's take a moment and quantify skewness, which is really a measure of the asymmetry of a distribution. From what we just described here, it is a measure of the asymmetry about the mean. It would be helpful to attach a number to this statistic to give one an idea of how serious the skewness may be. It is especially important in laboratory data analysis since in many such applications data is skewed, and this will dictate the type of statistical analysis that will be relevant or appropriate. Skewness has been quantified in several ways. A very simple measure of skewness is known as Pearson's Skewness coefficient (Joanes and Gill 1998) and is simply defined as the median subtracted from the mean and that difference is divided by the standard deviation or,

$$\text{Skewness} = \frac{\overline{X} - \text{median}}{S},$$ (2.1)

where $\overline{X}$ is the mean and S is the standard deviation. It is a simple measure of how the mean and median differ in standard deviation units. According to this formulation, if the skewness < -0.20, one has severe left skewness, and if skewness $> +0.20$ then

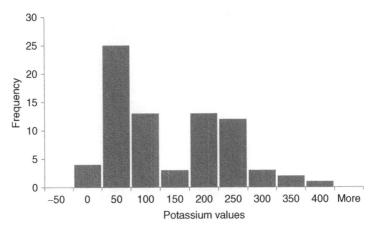

Figure 2.3   Distribution of 76 Potassium Values – Northeast Lab

**TABLE 2.2   Descriptive Statistics of 76 Potassium
Values – Northeast Lab**

Mean	121.974
Standard deviation	100.030
Standard error	11.474
Mode	35
N	76
Minimum	0
25th percentile	35
Median	92.5
75th percentile	200
Maximum	390

one has severe right skewness. The more complicated formula is called the adjusted Fisher–Pearson Standardized Moment Coefficient (SMC). This formula for skewness is the one that most statistical software packages use. The term "moment" in statistics refers to a measure relative to the center of the distribution. Its explanation can be quite involved, but for our sake we are measuring something relative to the center or mean. Thus, the Fisher–Pearson skewness formula is

$$\text{SMC} = \frac{n^2}{(n-1)(n-2)} \sum_{i=1}^{n} \left(\frac{X_i - \overline{X}}{S} \right)^3, \tag{2.2}$$

where n is the sample size, X_i is an individual data point in the distribution, $i = 1, \ldots, n$ and, as usual the sample mean is $\overline{X}$ with standard deviation, S.

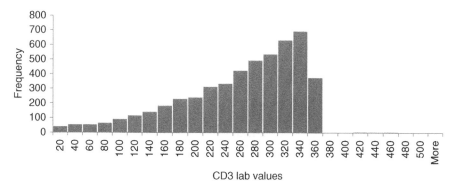

Figure 2.4 CD3 Laboratory Values from Cytometry Data

TABLE 2.3 Descriptive Statistics of 500 Left Skewed Data

Mean	249.107
Standard error	3.646
Median	271.509
Mode	337.000
Standard deviation	81.533
Range	337.908
Minimum	11.712
Maximum	349.620
N	500

According to this measure, we have the following skewness severity rule:

> If SMC > 1 moderate right skewness
>
> If SMC > 2 severe right skewness
>
> If SMC < − 1 moderate left skewness
>
> If SMC < − 2 severe left skewness.   (2.3)

We consider an example of these measures. We have potassium data (the X_i's) in Table 2.4. These measurements are in mEq/L or milliequivalents per liter. The summary statistics are given in Table 2.4. The normal range in the blood is 3.5–5.2 mEq/L. Clearly, we have some values skewed to the high side as seen in Figure 2.5a. We have overlaid a distribution curve to denote the direction of skewness. Note that according to formula (2.2), the SMC = 1.774, which is moderate to severe right skewness. The median of these values is four. According to the Pearson skewness formula or (2.1), we have Skewness = (5.44 − 4)/4.19 = 0.343, which indicates severe right skewness.

TABLE 2.4 Example of Skewed Data

i	X_i = Potassium Values	$Y_i = \ln(X_i)$
1	3.2	1.163
2	4.2	1.435
3	3.8	1.335
4	2.3	0.833
5	1.5	0.405
6	15.5	2.741
7	8.5	2.140
8	7.9	2.067
9	3.1	1.131
10	4.4	1.482
Mean	$\overline{X} = 5.44$	$\overline{Y} = 1.47$
Median	4	1.38
Variance	$S_X^2 = 17.53$	$S_Y^2 = 0.46$
Standard deviation	$S_X = 4.19$	$S_Y = 0.68$
Standard error of mean	$S_{\overline{X}} = 1.32$	$S_{\overline{Y}} = 0.22$

Often such data can be given a more normal or symmetric shape by using the natural log transformation denoted by the Y_i values in the last column of Table 2.4. Note the distribution of the Y values in the table and in Figure 2.5b. This is actually a log normal distribution. Here, the SMC = 0.433, or no skewness. Applying the Pearson formula, this data has a median of 1.38, and thus the skewness is $(1.47 - 1.38)/0.68 = 0.13$ or accordingly, no severe skewness. Thus, one can use the Pearson formula, which is easily calculated by hand, to get a good idea as to whether or not the data is severely skewed. Examining the graph of the frequency data should indicate visually if the calculation of skewness may be necessary. There are more formal statistical procedures for testing of normality or data symmetry. However, what is described here is certainly a good first step. As we'll discuss as we proceed through this chapter, we have ways of accommodating the analysis given the type of distribution we are dealing with.

The Normal Distribution is the most commonly used distribution in practice, and it can be characterized by symmetry and bell shape as shown in Figure 2.6 with equal central tendency values (mean = median = mode). Furthermore, the normal distribution takes on both negative and positive values, ranging from $-\infty$ to $+\infty$ (minus to plus infinity), and has little or no skewness.

Figure 2.6 describes different types of normal distributions. They differ in that the wider distribution (top panel) has a large variance or spread, and the more narrow distribution has smaller variance or spread as seen in the panels.

The Standard Normal (Gaussian) Distribution is often called the Z-distribution and Z-statistic (value) is defined by

$$Z = \frac{(Y - \text{mean})}{S}, \tag{2.4}$$

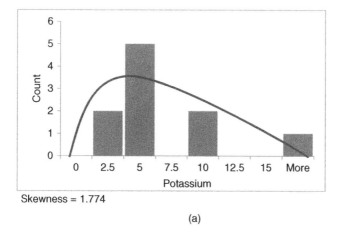

Skewness = 1.774

(a)

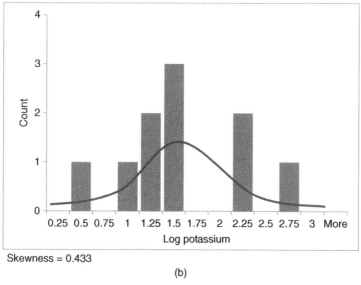

Skewness = 0.433

(b)

Figure 2.5 Distribution of (a) 10 Potassium Values Skewness = 1.774; (b) Log Transformed Potassium Values Skewness = 0.433

where Y is a variable of interest, normally distributed, and mean and S are the population mean and standard deviation, respectively.

The Standard Normal (Gaussian) Distribution has the same properties as the normal distribution and always has a mean value of 0 and a standard deviation value of 1; this is why converting the Gaussian Distribution into standard normal distribution is sometimes referred to as standardization.

For our BUN data, we estimate the population mean by the sample mean (=28.36) and the population S is estimated by the sample standard deviation (=7.889) as seen in Table 2.1.

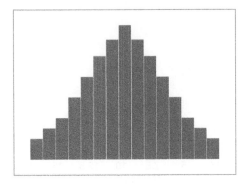

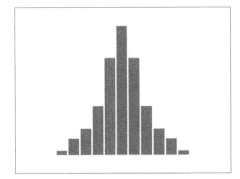 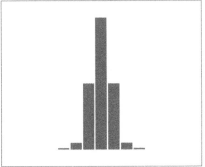

Figure 2.6 Types of Normal Distributions

So for each observation, Y_i, the standardized value is $Z = \dfrac{(Y_i - 28.36)}{7.889}$, for BUN values $i = 1, 2, \ldots, 116$.

For example, suppose in our sample of 116, one of the BUN values is 44. Then, its standardized value is

$$Z = \frac{(44 - 28.36)}{7.889} = 1.983. \tag{2.5}$$

Now, suppose in our sample of 116, one of the BUN values is 25. Then, its standardized value is

$$Z = \frac{(25 - 28.36)}{7.889} = -0.426. \tag{2.6}$$

Clearly, for values to the left of the mean (less than the mean) the Z-statistic will take on a negative value and for data values to the right of the mean (greater than the mean) the Z-statistic will be positive. This is one reason why we say the normal distribution can take on positive and negative values. The distribution of the standardized BUN values will have a mean of 0 and a standard deviation $= 1$.

Now let's look at the standardized BUN value distribution in Figure 2.7. The descriptive summary of the data is found in Table 2.5.

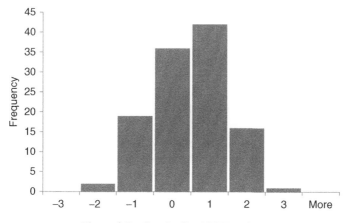

Figure 2.7   Standardized BUN Values

TABLE 2.5   Descriptive Statistics of Standardized BUN Values

Mean	0.0
Standard deviation	1
Standard error	0.093
N	116
Minimum	−2.669
25th quantile	−0.507
Median	0.011
75th quantile	0.643
Maximum	2.928

As one can identify from the descriptive statistics in the figure, this distribution has a mean of 0 and S of 1.0, and the values range from −2.67 to +2.93.

2.2   CONFIDENCE INTERVALS (CI)

Up to now, we have been discussing point estimates of the population values. For example, the sample mean is a point estimate of the population mean. It is just a single value. Similarly, the sample variance is the point estimate of the population variance. It is just one value. A Confidence Interval (CI) is an "interval estimate" of a population value. It tells us how certain we can be that a parameter, such as the population mean, lies between two limits or values. It takes the form:

$$(\text{lower limit, upper limit}). \qquad (2.7)$$

For example, let's consider our BUN laboratory data again. There are 116 observations ranging from 7.3 to 51. Then, based on this sample, again we have the following results:

A 95% CI for the true population BUN mean is (26.92, 29.80).

- Recall the BUN data; we have 116 BUN values from a Northeastern Laboratory. The distribution is as in Figure 2.1.

Based on this sample, we can be certain with 95% probability that this interval contains the true population mean or the population mean lies somewhere between 26.92 and 29.80. Most statistical computer programs generate CIs routinely. Most are too complex to do by hand. However, in the following section we demonstrate in a particular parameter how one can calculate it by hand.

2.2.1　Confidence Interval (CI) for the Population Mean – The t-Distribution

Remember that the CI always refers to a population value, but uses the sample values in its calculations. Thus, in our case, to calculate the CI for a population mean, the CI is a function of the sample mean, $\bar{x}$, the sample standard error (SE), and a value known as a t-value. The lower limit is $\bar{x} - t * \text{SE}$. The upper limit is $\bar{x} + t * \text{SE}$. The t-value is based on $n - 1$ degrees of freedom (df), where n is the sample size and the level of confidence desired (typically 90%, 95%, and 99%). It is generated by the computer and is found in most statistics text books.

Example of Normal Data　For the sake of demonstration, the following is a normal sample of laboratory data (24 observations)

$$0.5, 2.5, 3.45, 3.5, 4.5, 4.51, 1.5, 2.5, 3.5, 3.5, 4.5, 5.5,$$

$$1.5, 2.5, 3.5, 3.5, 4.5, 5.52, 2.5, 2.56, 3.5, 3.5, 4.5, 6.5$$

The descriptive statistics are given in Table 2.6. Commonly used confidence levels (t-critical value) for $n = 24$ (df $= n - 1 = 23$) as in our laboratory data, which are found in t-tables of statistics books are as follows:

Level of Confidence	t – Critical Value	Confidence Interval	
90	1.714	(3.018, 3.986)	
95	2.069	(2.918, 4.085)	
99	2.087	(2.913, 4.091)	(2.8)

Thus, the 95% CI for the population mean based on the t-distribution ($n = 24$) using the laboratory data values from Figure 2.8 (mean$(\bar{X}) = 3.502$, SE $= 0.282$) and

TABLE 2.6   Example of Normal Data ($n = 24$)

Descriptive Statistics	
Mean	3.502
Standard error	0.282
Median	3.5
Mode	3.5
Standard deviation	1.383
Skewness	0.004
Minimum	0.5
Maximum	6.5
N	24

$t = 2.069$ from above is given by

$$(\bar{x} - t * \mathrm{SE}, \bar{x} + t * \mathrm{SE}) = (3.502 - 2.069(0.282), 3.502 - 2.069(0.282))$$

$$= (2.918, 4.085). \tag{2.9}$$

Thus, we can be 95% certain that the true population mean for our laboratory data lies between 2.918 and 4.085 or alternatively, a 95% CI for the true population laboratory mean is (2.918, 4.085). If we wanted a 99% CI using this sample, the mean and SE are the same, but the t-value would be 2.087 and the CI $= (2.913, 4.091)$. Keep in mind that this is just one set of data of this kind. The CI depends on the sample size, the mean, the SE, and the t-value. If any of these values change, then so would the CI. As is seen in (2.8), we have a different set of intervals for this same data, but with varying levels of confidence.

Thus, in summary, note that in general, the higher the level of confidence in (2.8), the wider the interval, that is, the larger the t-value. The mean ($\overline{X}$) and SE are derived from the sample. The t-value depends on the sample size, n, and the confidence you wish to determine. Graphically, one can see from Figure 2.8 that if the mean is in the center of the distribution, then the vertical lines represent the width of the CIs for varying levels of confidence.

2.2.2   Confidence Interval for the Variance and Standard Deviation

One can put a CI on any parameter of interest (e.g., standard deviation). However, the calculations can become quite complex because they may depend on other types of distributions as the CI for the mean depends on the t-distribution above. As an example, the CI for the standard deviation depends on a different theoretical distribution called the chi-square (χ^2) distribution.

As we can see later, the calculations are quite involved. However, strictly for the sake of demonstration, we will show that given the standard deviation in Table 2.6, 1.383, from our 24 observations of laboratory data, we can get a CI on the population standard deviation.

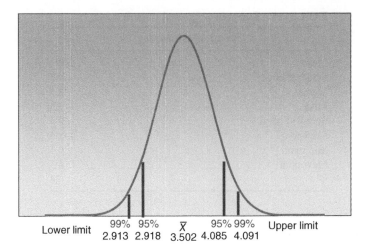

99% 95% $\bar{X}$ 95% 99%   Upper limit
Lower limit 2.913 2.918 3.502 4.085 4.091

Figure 2.8 Example of Normal Data ($n = 24$)

This is a measure of the precision of the laboratory sampling or readings. The CI for the variance is

$$\frac{(n-1)\,s^2}{\chi^2_{\alpha/2}} \leq \sigma^2 \leq \frac{(n-1)\,s^2}{\chi^2_{\left(1-\alpha/2\right)}},\tag{2.10}$$

where the chi-squared test statistic

$$\chi^2 = \frac{(\mathrm{df})\,s^2}{\sigma_0^2},\tag{2.11}$$

where n and s^2 are sample size and sample variance, respectively, and

$\sigma_0^2 = $ hypothesized population variance, and df(degrees of freedom) $= n - 1$.

The CI for the population standard deviation is

$$\sqrt{\frac{(n-1)s^2}{\chi^2_{\alpha/2}}} \leq \sqrt{\sigma^2} \leq \sqrt{\frac{(n-1)s^2}{\chi^2_{\left(1-\alpha/2\right)}}}.\tag{2.12}$$

From the results of normal laboratory data, note the standard deviation is equal to 1.383. Formula (2.12) gives the upper and lower limits of the CI of the standard deviation, which is quite tedious to calculate, but is done routinely by the computer. The limits are (1.075, 1.934).

So we can interpret the CI as saying we can be 95% certain that the interval, (1.075, 1.934) covers the true population standard deviation, that is, the interval, (1.075, 1.934) is a 95% CI for the population standard deviation based on our sample.

Thus, in Chapter 1 and this chapter so far we have covered a number of descriptive statistical terms we use in laboratory data analysis. Some of the point estimates we have discussed are the mean, variance, standard deviation, SE of the mean, and coefficient of variation (RSD, relative standard deviation). For the interval estimates, we have discussed CIs.

2.3 INFERENTIAL STATISTICS – HYPOTHESIS TESTING

Now, we would like to study common terminology associated with formal statistical hypothesis testing. Next, we cover types of errors in hypothesis testing, and then we learn to interpret the meaning of p-values.

Hypothesis Testing A hypothesis is a belief that may or may not be true but is conditionally assumed true until new evidence (collected data) suggests otherwise. For example, we might assume before performing the experiment that the population means from two different laboratory samples collected under different conditions or methods (condition 1 and condition 2) are the same. Our goal may be in fact to show that the two methods are not the same with respect to their mean outcome. We may wish to show that they are in fact different. The null hypothesis would be that they are the same (status quo). Let's assume that μ_1 is the population mean under condition 1 and μ_2 is the population mean under condition 2. We denote this null hypothesis as H_0

$$H_0 : \mu_1 = \mu_2. \tag{2.13}$$

The research question or goal is that they are different; we denote this as the alternative hypothesis or as H_1, that is,

$$H_1 : \mu_1 \neq \mu_2. \tag{2.14}$$

There are four steps in Hypothesis Testing:

1. Set up the hypotheses.
2. Determine the significance level.
3. Collect evidence and compute the p-value.
4. Make a decision.

We will show examples of these steps and how hypothesis testing is actually performed. However, let's first discuss some rules and definitions for such. Suppose we used a rule to make decisions, but were the decisions correct? There are two types of errors that we try to minimize: (1) Type I error and (2) Type II error. Recall from Chapter 1 that we work with the sample and generalize the results to the population.

Type I error: Given our aforementioned example, a test from the sample may lead to a conclusion that the population means are significantly different when, in fact, they are not. In other words, one rejects H_0 in favor of H_1. That is to say, you conclude the two means to be different when in fact they are not. The probability of a type I error is denoted by α $(0 < \alpha < 1)$. The null hypothesis is always no difference (no effect, status quo) versus the alternative hypothesis that there is a difference (i.e., an effect). If the experiment from the sample indicated that there was a difference when, in fact in reality, in the population there was not a difference, then a type I error has been made.

Type II error: We accept H_0 from the sample when in fact H_1 is true (i.e., H_0 is not true). A test may lead to a conclusion that the two means are not significantly different when, in fact, they are. The probability of a type II error is denoted by $\beta(0 < \beta < 1)$. Using the same aforementioned example, in this case if the experiment indicated that there was no difference, when in fact in reality, in the population there was a difference, then a type II error has been made.

Power: The probability $(1 - \beta)$ is called the power of a statistical test. That is the probability of making a decision of a difference when in fact in the population there was a difference. In other words, once we reject the null hypothesis and declare an effect the power is the probability of detecting an effect or change when it exists, that is, it is the confidence we have in the effect we observed. Thus, one can see that in hypothesis testing one wants to minimize both errors (α and β) and maximize the power.

Table 2.7 is a schematic of the decision process and types of errors one can make. This type of table is found in most statistics texts.

Comparing α and the p-*Value* Now we have all the elements of hypothesis testing as given in the four steps above.

In most statistical hypothesis tests, we do the following:

1. State the hypotheses as in (2.13) and (2.14).
2. Determine the significance level, that is, specify α, $(0 \leq \alpha \leq 1)$, the probability of a type I error, which is determined by the experimenter before collecting data (running the experiment, trial, etc.). Note that α is called the significance level, which is usually set very low at about 0.05 in most of applications.
3. Collect the evidence from the experiment, that is, collect the data. The p-value is calculated from the collected data, $(0 \leq p \leq 1)$. It is a measure of how much

TABLE 2.7 The Decision Process in Hypothesis Testing

Decision (Sample)	Actual (Population)	
H_o	H_o True	H_o Not True
H_o true	Correct $(1 - \alpha)$	Type II error (β)
H_o not true	Type I error (α)	Correct $(1 - \beta)$

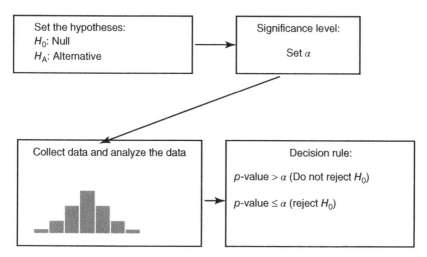

Figure 2.9   Steps in a Statistical Hypothesis Test

the data support the null hypothesis. Some refer to the p-value as the amount of evidence against the null hypothesis. In other words, the smaller the p-value, the more evidence against the null hypothesis.

4. Make a decision. Reject or do not reject H_0. In general, we reject the null hypothesis if $p \le \alpha$ where α is usually equal to 0.05 and fail to reject the null hypothesis if $p > \alpha$.

These four steps are also diagrammed in Figure 2.9.

2.3.1   *t*-Test for Means

In this section, we learn to construct what are called lower tailed, upper tailed, and two-tailed hypothesis tests by using the z-test and the t-test. We identify the appropriate test statistic and hypothesis test to address these issues. Our goal in this section is to determine how likely the hypothesized value is from the data we sampled. We will concentrate first on a single sample. Suppose we want to conduct an experiment and collect data on a laboratory value from subjects, and we know from historical evidence that the expected value of the population mean, μ, of the laboratory result takes on the value, μ_0. We are, in fact, going to take a sample from a sick group of subjects and want to show that their laboratory result will not be the expected value in this case. So our null hypothesis is

$$H_0 : \mu = \mu_0. \tag{2.15}$$

Our research question or alternative hypothesis that we want to prove is that

$$H_1 : \mu \ne \mu_0. \tag{2.16}$$

If $\bar{x}$ is the observed mean of the sample data, the question is how far apart is $\overline{X}$ (sample mean) from the hypothesized mean, μ_0, or does the data yield a sample mean much different from the expected, μ_0? The benchmark that we use to measure the distance between the observed sample and the hypothesized null value is the SE.

In other words, is $[\overline{X}(\text{sample mean}) - \mu_0]$ big or small in SE units? How do we determine that?

The test statistic follows what is known as a t-distribution, which we construct here. For example, let $\bar{x}$ = sample mean, μ_0 be the population mean under the null hypothesis and SE be the sample SE, then one computes

$$t_{\text{calc}} = \frac{(\overline{X} - \mu_0)}{\text{SE}}. \tag{2.17}$$

If the calculated t_{calc} or t is close to zero, then there is evidence in the sample for the null hypothesis. If t is much different from zero, then there is little evidence in the sample for the null hypothesized value of the population mean, that is, $\overline{X}$ and μ_0 differ or the calculated mean we observe in the sample, $\overline{X}$, is different from what we expect in the population, μ_0. Thus, the sample tells us that we are different from the hypothesized mean and we may reject H_0. Let's do a computational example according to the four steps in (2.15).

Computation Example – One Sample t-Test From the data in the BUN example, Table 2.1, we know that the BUN range for this diseased population is from 7.3 to 51.5. In general, the normal healthy range of BUN is considered to be 7–20 mg/dL. Suppose further that historically we have observed a mean BUN value of 28 in a similar diseased group. Suppose the group noted in Table 2.1 was treated and it is hoped that the mean value is different from $\mu_0 = 28$. Let's follow the steps in (2.15) to conduct our hypothesis test.

1. State the hypotheses:
   H_0: population mean(μ_0) = 28 versus H_1: population mean(μ_0) $\neq$ 28, or similarly,
   H_0: $\mu_0 = 28$ versus H_1: $\mu_0 \neq 28$.
2. Determine the significance level, that is, specify $\alpha = 0.05$, level of significance. Note that here H_1 implies the true population mean is either greater than 28 or less than 28.
3. Collect the evidence from the experiment, that is, collect the data. The p-value is calculated from the collected data. It is not straightforward to calculate and is usually done by computer.

From Table 2.1, $\overline{X} = 28.36$ and the SE = 0.73. The test statistic from the sample data according to formula 2.17:

$$t_{\text{calc}} = \frac{28.36 - 28}{0.736} = 0.489. \tag{2.18}$$

or $p = 0.6256$ (which is >0.05). The t-critical value corresponding to the two-sided probability of 0.05 for $n = 116$ $(117 - 1)$ is found in t-tables of statistic books is 1.981. The calculated t-test statistic from the sample data is 0.489. This test statistic is less than the t-critical value of 1.981. The p-value is calculated from the t-value of 0.489. It is not straightforward to calculate and is usually done by computer.

4. Make a decision. Reject or do not reject H_0. In general, we reject the null hypothesis if $p \leq \alpha$ and do not reject H_0 if $p > \alpha$. Clearly, we do not reject H_0 since $p = 0.6256 > \alpha = 0.05$. Note $t_{calc} = 0.489$ is well within the expected distribution under the null hypothesis which is seen in Figure 2.8. Certainly, the mean BUN based on our sample is not statistically different from 28.

So Where Does p Come From? Conceptually, the p-value is determined by the t-statistic for this case. By focusing on equation (2.17), we note that as t increases, the sample mean moves further away from the expected population mean and we have less evidence in favor of the null hypothesis, so the p-value decreases. As t decreases, the sample mean moves closer to the population mean and we have more evidence in favor of the null hypothesis, so the p-value increases. Setting $\alpha = 0.05$, if $p < \alpha = 0.05$, then one can think of interpreting the p-value as telling us that there is less than a 5% chance that the null is true. We thus reject it. If $p > \alpha$, then there is greater than a 5% chance that the null hypothesis is true, and we do not reject it. Note that the α is set very low to minimize making the type I error or rejecting a null that may in fact be true. In our aforementioned example, $p = 0.62$ or there is certainly a pretty high probability that the null is true and we do not reject it based on our sample results.

In this section thus far, we have demonstrated a two-sided test. This is noted by the alternative hypothesis where $H_1 : \mu \neq \mu_0$. As mentioned earlier, this could mean that the true population mean could be greater than the value of μ_0 or less than the value of μ_0. Suppose our research question or alternative hypothesis is stated a bit differently.

Explanation of One-Sided Tests Using our aforementioned data example, suppose one wished to show that the data supported a value greater than 28, that is, we wished to test: H_0 : population mean $(\mu_0) \leq 28$ versus H_1 : population mean $(\mu_0) > 28$, this is called an upper tail test. Without going through all the details, the p-value generated by the computer would be $p = 0.311$. Conversely, if one wished to show that the data supported a value less than 28, that is, we wished to test: H_0 : population mean$(\mu_0) \geq 28$ versus H_1 : population mean$(\mu_0) < 28$, this is called a lower tail test. The p-value generated by the computer would be $p = 0.689$. The point being that no matter what the null or alternative hypotheses, the p-value generated will always be interpreted as we stated. That is, we reject the null hypothesis if $p \leq \alpha$ where α is usually equal to 0.05, and fail to reject the null hypothesis if $p > \alpha$. This will be the rule throughout this chapter and several of the chapters to follow. Also, note that α or the type I error can be set at any value an experimenter wishes. The value 0.05 is the norm for many statistical applications. However, α levels of 0.01, 0.025, and 0.10 are not uncommon, depending on the application. Also, note in our example here that the alternative hypothesis is the research question we are interested in. This is what we are trying to prove.

Known Standard Deviation: the z-Test The Z-statistic follows a standard normal distribution with mean zero and standard deviation one. For example, again let $\overline{X}$ equal to the mean from the sample and μ_0 be the hypothesized population mean under null hypothesis and SE be the known population SE, then one computes

$$Z = \frac{(\overline{X} - \mu_0)}{\text{SE}}. \tag{2.19}$$

Like the *t*-statistic, if z is close to zero, then there is evidence in the sample for the null hypothesis. If z is much different from zero, then there is little evidence in the sample for the null hypothesized value of the population mean.

Which Test Statistic Should We Use? We use the z-test for the mean when we assume a random sample is drawn from the population normally distributed with a known population standard deviation. However, in most practical experiments the population standard deviation is not known. Therefore, the *t*-test is more commonly used for inference about means. One needs to keep in mind, as we've mentioned earlier in the chapter, that we cannot always assume that the data is normally distributed (symmetric). In that case, one uses a nonparametric test for our single sample called the Wilcoxon signed rank test to make inference about the mean. We don't go into the details here, as it can become very mathematically cumbersome. However, for our aforementioned example, testing $H_0 : \mu = 28$ versus $H_1 : \mu \neq 28$, the Wilcoxon statistic generates a *p*-value $= 0.573$, which is nonsignificant and certainly consistent with our earlier decision of not rejecting the null hypothesis.

2.3.2 Test for Variation: Coefficient of Variation (CV)

Suppose we approach our hypothesis testing strategy a little differently, and we wish to hypothesize that the variance (Standard Deviation $= \sigma_0$) takes on a certain value. This is like a measure of the precision of the experiment, that is, test $H_0 : \sigma_0 = 5.6$ versus $H_a : \sigma_0 \neq 5.6$ at $\alpha = 0.05$. Note here if the population mean, μ_0, is 28, then the motivation for $\sigma_0 = 5.6$ is that $(5.6/28.0) \times 100 = 20$, that is, using $\sigma_0 = 5.6$ yields a CV of at most 20%. That is to say in a laboratory setting that we wish to test that the precision of the experiment is at most 20%. Then based on the CV, a mean of 28 and a S of 5.6 will give us that precision. We will use our data example to perform a hypothesis test of the Standard Deviation equal to 5.6. The summary statistics are provided in Table 2.1. The actual sample CV $= (7.889/28.36) \times 100 = 27.8\%$. The actual *p*-value for testing that the CV in this case is 20% or less is about $p = 0.001$. Since $p < 0.05$, we reject $H_0 : \sigma_0 = 5.6$ or there is very little chance that that the population S $= 5.6$ or the CV $\leq 20\%$. Thus, we can conclude the experiment or process is not very precise. Therefore, one sees

TABLE 2.8   Test Statistics for Mean Difference: (Frozen Serum – fresh Serum) – Paired t-Test

Fresh serum mean	222.794
Frozen serum mean	222.608
Mean difference	−0.1865
Standard deviation	3.832
Standard error	0.465
df	67
t-Test statistic	0.400
p-Value	0.689
t-Critical	1.997

df: (degrees of freedom) the number of independent observations minus 1.

that the rules for hypothesis testing do not change. Once the null and alternative hypotheses are stated, and if $p \leq 0.05$, then we reject the null.

2.3.3   Two-Sample Test of the Population Means

t-Test Statistic for Dependent Samples   The paired sample t-test is sometimes used in the laboratory for method comparison. As an example, suppose we have total cholesterol (TCL) data from several individuals. Their serum was drawn, and the TCL was determined from this fresh serum. The serum was then frozen and the TCL was read from the frozen serum. The paired test is used to determine if the mean difference in TCL from a fresh sample compared to a frozen sample is significant. Such data was recorded from 68 individuals and the summary statistics are provided in Table 2.8. The first four data points look as follows:

Subject	Frozen TCL	Fresh TCL	Difference
1	227.3	223.1	4.2
2	196.4	198.2	−1.8
3	200.3	202.6	−2.3
4	176.2	170.0	6.2

Data as such is recorded on the remaining 68 subjects. Therefore, we have 68 paired differences. If there were no difference on average between the fresh and frozen samples, then we would expect the average (mean) difference (−0.1865 in the aforementioned table) to be close to zero. Thus our hypotheses are as follows:

$$H_0 : \mu_{\text{Difference}} = 0 \text{ versus } H_a : \mu_{\text{Difference}} \neq 0. \tag{2.20}$$

We use this t-test to determine whether two means from paired samples differ. Defining the sample mean difference as $\overline{X}_{\text{Difference}}$ and SE as, $SE_{\text{Difference}}$, our paired t-statistic now takes the form:

$$t_{\text{calc}} = \frac{(\overline{X}_{\text{Difference}}) - 0}{SE_{\text{Difference}}} = \frac{-0.1865}{0.465} = 0.400, \qquad (2.21)$$

where 0 is the hypothesized difference. This statistic is done computationally exactly as equation (2.17) with the sample mean and SE calculations performed on the differences or last column in the aforementioned example. The interpretation of the t-statistic is exactly as we discussed in our previous example. That is, if on average the change in Frozen–Fresh is large, then t is large and p is small (i.e., $p \leq 0.05$) and we reject the null hypothesis of no difference. See the data summary on Table 2.8. Note $p = 0.689$, or we do not reject $H_0 : \mu_{\text{Difference}} = 0$ or we determine there is no difference between the Fresh and Frozen serum on average. Figure 2.10 demonstrates graphically how the Fresh and Frozen sera compare. The horizontal axis is the average of the two for each of the 68 data pairs, that is, (Frozen + Fresh)/2 and the vertical axis is the difference between the Frozen and Fresh value (Frozen–Fresh), for each individual. Note that how the scattering of point differences is centered around zero, which supports our conclusion of not rejecting the null, that is, there is no difference between fresh and frozen serum. As we mentioned earlier, this paired difference strategy is one of the strategies often used for laboratory method comparison and method validation, which we will discuss in Chapter 3. We will discuss other strategies as well. For example, our fresh and frozen samples could have just as easily been two different laboratory methods for measuring an analyte, and we want to know if the methods are comparable.

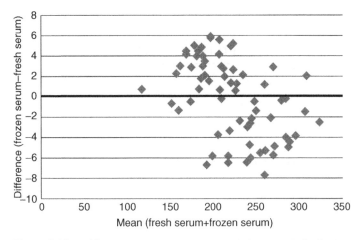

Figure 2.10   Difference: (Frozen Serum – Fresh Serum) – Paired t-Test

Two-Sample t-Test Now we wish to compare the means from two independent populations assuming independent observations, normally distributed sample means for each population, and equal variances for each treatment.

For example, we might want to know if the mean BUN value in one group is the same as the mean BUN value in another group. We take a sample from each group and determine if their means are the same statistically. These are not paired observations as in our aforementioned *t*-test dependent example, where we got two readings per individual and test the mean difference. In this case, there are two different sets or samples of individuals and we will compute a mean BUN value for each sample. This is like comparing two different treatments, interventions, or laboratory techniques with the goal of determining if they are different. Let μ_{BUN1} be the mean value of the population from which the first group was drawn and μ_{BUN2} be the mean value of the population from which the second group was drawn. Then, in terms of a hypothesis test, we have

$$H_0 : \mu_{BUN1} = \mu_{BUN2} \quad \text{versus}$$
$$H_a : \mu_{BUN1} \neq \mu_{BUN2}. \tag{2.22}$$

Stated another way, we could have

$$H_0 : \mu_{BUN1} - \mu_{BUN2} = 0 \quad \text{versus}$$
$$H_a : \mu_{BUN1} - \mu_{BUN2} \neq 0. \tag{2.23}$$

Let us denote the observed sample means as $\overline{X}_{BUN1}$ and $\overline{X}_{BUN2}$. The SE of the difference is denoted as $SE_{BUN1-BUN2}$.

Then we use the *t*-test to test whether two means are different. In this case, it takes the form

$$t = \frac{(\overline{X}_{BUN1} - \overline{X}_{BUN2}) - 0}{SE_{BUN1-BUN2}}, \tag{2.24}$$

where 0 in this equation is the hypothesized difference. Equation (2.24) is computationally the same as equation (2.17) where the $\overline{X}$ in equation (2.17) is now the difference between $\overline{X}_{BUN1}$ and $\overline{X}_{BUN2}$ in equation (2.24) and the SE in (2.17) is the $SE_{BUN1-BUN2}$ in (2.24). This $SE_{BUN1-BUN2}$ is computationally a bit more involved and easily done by the computer, but also can be done by hand. For one who wishes to do so, we present the formula as follows:

$$SE_{BUN1-BUN2} = \sqrt{S_P^2 \left(\frac{1}{n_1} + \frac{1}{n_2} \right)}, \tag{2.25}$$

$$S_P^2 = \frac{(n_1 - 1)S_1^2 + (n_2 - 1)S_2^2}{n_1 + n_2 - 2}, \tag{2.26}$$

TABLE 2.9 Test Statistics and p-Values for t-Test for Two Groups Assuming Equal Variances

	BUN1	BUN2	Mean Difference
Mean	27.641	33.672	6.031
Variance	57.033	60.503	—
Observations	91	25	—
Pooled variance	57.764	—	—
Hypothesized mean difference	0	—	—
df	114	—	—
t-Statistic	3.514	—	—
p-Value	0.0006	—	—
t-Critical value	1.981	—	—

where S_p^2, S_1^2, S_2^2, n_1, and n_2 are the pooled variance of both BUN samples, variance of the first BUN sample, variance of the second BUN sample, the first BUN sample size, and the second BUN sample size, respectively.

Note that from Table 2.9, we have $n_1 = 91$ and $n_2 = 25$. Note the sample size need not be the same in both groups. Also S_1^2 is the variance of the first BUN sample or 57.033 and S_2^2 is the variance of the second BUN sample or 60.503. Thus, doing the calculations the $S_p^2 = 57.764$,

$$ SE = \sqrt[2]{S_p^2 \left(\frac{1}{n_1} + \frac{1}{n_2} \right)} = \sqrt[2]{57.764 \left(\frac{1}{91} + \frac{1}{25} \right)} = 1.716 $$

and

$$ t = \frac{(33.672 - 27.641) - 0}{1.716} = 3.514. \tag{2.27} $$

Figure 2.11(a) shows the distribution of this t-statistic. If the null hypothesis were true, then the sample mean difference, $\overline{X}_{BUN1}$ and $\overline{X}_{BUN2}$ is 0 or the t-value or statistic above is small and close to the center of Figure 2.11(a). If the null were not true, then the t-value is large and could be to the right or left of the center and possibly beyond the critical regions, which denotes a statistical difference between the two means. The area under the distribution curve is always equal to one. Commonly used confidence levels (t-value) for $n = 114$, which are found in t-tables of statistics books are as follows:

Level of Confidence	t-Value
90	1.658
95	1.981
99	2.619

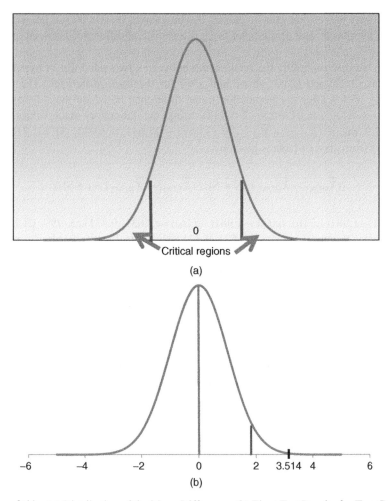

Figure 2.11   (a) Distribution of the Mean Difference; (b) The t-Test Results for Two Groups

As before, if the difference in sample means between two groups is large, then t is large and p is small (i.e., $p \leq 0.05$) and we reject the null hypothesis of no difference. There were 91 observations from Group 1 and 25 observations from Group 2. Please see the summary data in Table 2.9 and Figure 2.11(b). The t-value from Table 2.9 is 3.514, which represents the black vertical line in Figure 2.11(b). This line is beyond the critical cutoff of ± 1.98 noted in Table 2.9, which cuts off the 0.025 region from the two extreme tails of Figure 2.11, or a total area under the distribution curve of 0.05. This $t = 3.514$ is beyond the 1.98 cutoff, and yields an area or p-value of $p = 0.0003$. However, our alternative was $H_a : \mu_{\text{BUN1}} - \mu_{\text{BUN2}} \neq 0$. Then, we consider that this t-value could be positive or negative, that is, fall in either critical region. So we have a two-tailed test that we discussed earlier, and the p-value is double to $p = 0.0006$, which is why we see that value in Table 2.9. Thus, since $p = 0.0006 < 0.05$, we reject the null hypothesis and declare that there is a statistical difference between the two means.

Then we discuss briefly the relationship between a two-sided test of hypothesis, equations (2.22) and (2.23) above and a CI for the mean difference. The corresponding 95% CI for the population mean difference based on the t-distribution (df $= n_1 + n_2 - 2 = 91 + 25 - 2 = 114$) using the laboratory data values from Table 2.9 mean $(\overline{X}_{BUN1} - \overline{X}_{BUN2}) = (33.672 - 27.641) = 6.031$, SE $= 1.716$ and $t = 1.981$ from above t-table is given by

$$((\overline{X}_{BUN1} - \overline{X}_{BUN2}) - t * \text{SE}, \ (\overline{X}_{BUN1} - \overline{X}_{BUN2}) + t * \text{SE}) \qquad (2.28)$$

$= (6.031 - 1.981(1.716), 6.031 - 1.981(1.716)) = (2.631, 9.431)$. A 95% CI for the difference in population means is 2.631–9.341 BUN units. In other words, we are 95% confident that the population mean difference in the BUN value will be between 2.631 and 9.341. Note that the null value for the test is 0, which is not covered by this CI, that is, 0 lies outside the 95% CI of the plausible mean difference for the population means based on this sample. Therefore, the mean difference is significantly different from the population mean difference value of 0.

In general, the CI is appropriate when we make estimation about the parameter of interest, the mean difference here in the example above. However, if we make a conclusion about a specific hypothesis test, then the hypothesis testing is more appropriate. Note that both approaches may be used in many situations.

Summary Results for Our BUN Data   We reject H_0. The two means are not equal at the 0.05 level since our p-value or likelihood of our actual result (sample mean difference $= 6.03$) is $p = 0.0006$ if the null hypothesis were true. We did reject the null hypothesis of no difference. As we mentioned in Section 2.3, when one rejects the null hypotheses, one can also examine the power of the procedure, that is, the probability of having correctly rejected a false null hypothesis. Like the p-value, this is not computationally straightforward and is done using a computer. For the sake of completeness for this example, the power was very high (0.99) at $\alpha = 0.05$ or the probability of detecting a true difference for this sample is very high. So we conclude the mean BUN values of Groups 1 and 2 are in fact different.

2.3.4   One-Way Analysis of Variance (ANOVA)

Suppose we want to compare three or more means, for example, k means where $k \geq 3$. The null and alternative hypotheses take on the form,

$H_0 : \mu_1 = \mu_2 =, \ \ldots, \ = \mu_k$ versus

$H_A :$ at least one μ_i not equal to one other $\mu_j, i \neq j, i, j = 1, \ldots, k.$ \qquad (2.29)

One way analysis of variance is called ANOVA because we are actually comparing the variation within groups to the variation between groups and translating that inference to the means comparisons. The test statistic for ANOVA is called the F-statistic and is defined as

$$F = \frac{\text{Variation between groups means}}{\text{Variaion within groups}} = \frac{\text{MST}r}{\text{MSE}}. \tag{2.30}$$

Theoretically, if one can show a much greater variation between groups than within groups, then the group means will have a greater separation and be different from one another. Clearly, in this F-statistic if the numerator is much greater than the denominator, then it will be the case. In one-way ANOVA, we have two degrees of freedom. The first one is called numerator degrees of freedom (number of groups $-$ 1) and the second one is called the denominator degrees of freedom (total sample size $-$ number of groups). The shape of the F-distribution depends on the degrees of freedom. Variation among group means is measured by the sum of squares between groups divided by the corresponding degrees of freedom; this quantity is called mean square for treatments (MSTr). The sum of squares between groups (SSBG) is computationally a bit more involved and easily done by the computer, but also can be done by hand. For one who wishes to do so, we present the formula as follows:

$$\text{SSBG} = n_1(\bar{x}_1 - \bar{x})^2 + n_2(\bar{x}_2 - \bar{x})^2 + \cdots + n_k(\bar{x}_k - \bar{x})^2, \tag{2.31}$$

where n_1 is the sample size of group 1, $\bar{x}_1$ is the mean level for group 1, n_k is the sample size of group k, $\bar{x}_k$ is mean level for group k and $\bar{x}$ is the overall mean. Similarly, the variation between group means is measured by the sum of squares within groups (SSWG) divided by the corresponding degrees of freedom; this quantity is also called the mean square error (MSE). The SSWG is computationally a bit more involved and easily done by the computer, but also can be done by hand. For one who wishes to do so, we present the formula as follows:

$$\text{SSWG} = (n_1 - 1)s_1^2 + (n_2 - 1)s_2^2 + \cdots + (n_k - 1)s_k^2, \tag{2.32}$$

where s_1^2 is the estimate of the variance for group 1 and s_k^2 is the estimate of the variance for group k. For our purpose, here we need not go into the detail of this translation demonstration. Owen and Siakaluk (2011) give a good example of such. Our purpose is to explain the concept and example of ANOVA for inference on three or more means. Suppose we have three effects, A, B, and C, and we wish to determine if any mean effect, μ (e.g., reduction in blood pressure, tumor response, and laboratory result) of one treatment is significantly different from the other(s).

Our null hypothesis is H_0: $\mu_A = \mu_B = \mu_C$ versus H_a: at least one of these mean effects is different from one other where A, B, and C are group variables. If the null hypothesis is true, we expect the F-statistic (ratio of the MSTr and MSE) to be close to 1. If one of these mean's effect is greatly different from another, we expect the ratio to be much larger than 1. p-values of the F-statistic are generated from (2.30) as an area under the F-distribution curve.

One-Way ANOVA Example   Suppose we do in fact have three treatment groups we call A, B, and C, and we wish to compare them with respect to their mean HDL levels. Our null hypothesis is H_0: $\mu_A = \mu_B = \mu_C$ versus H_a: at least one of these mean effects is different from one another. The three treatment groups we call A, B, and C are individuals with three different exercise regimens per week. Group A exercises less than 50 minutes per week, group B exercises 50–150 minutes per week, and group C exercises more than 150 minutes per week. The summary and test statistics for the three exercise groups are in Table 2.10A and 2.10B and Figure 2.12.

TABLE 2.10A Summary Statistics for the Three Exercise Groups

	A: < 50	B: 50–150	C: > 150
Mean	47.72	46.036	44.829
Standard error	1.772	1.217	1.843
Median	48.760	45.920	44.94192
Standard deviation	10.780	7.600	8.0311
Range	46.439	29.270	26.978
Minimum	22.456	30.201	31.952
Maximum	68.895	59.471	58.929
N	37	39	19

TABLE 2.10B Test Statistics for the Three Exercise Groups

Source of Variation	SS	df	MS	F	p-Value	F-crit
Between groups	116.296	2	58.148	0.7095	0.4946	3.095
Within groups	7539.893	92	81.955	—	—	—
Total	7656.189	94		—	—	—

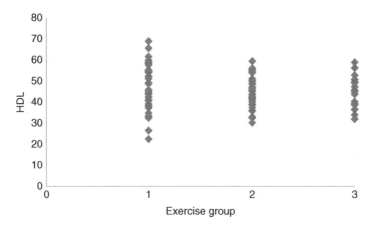

Figure 2.12   Distribution of Data for Each Group

We describe the summary results comparing the mean HDL levels for the three exercise groups. The ANOVA was used to test the simultaneous equality of all three means. Based on our results, it appears that the exercise regimen made no difference with respect to mean HDL level attained ($p = 0.4946$). The calculated value of $F = 3.095$ yields an area under the F-curve or p-value of $p = 0.4946$. In Figure 2.12, the distribution of points appears to be similar. In Table 2.10A, the three means appear to be similar as well, that is, the HDL readings for groups A, B, and C are 47.72, 46.04, and 44.83, respectively. Thus, our p-value of 0.4946, which certainly is >0.05, makes sense and we do not reject the null hypothesis of equal means. If significance was found, that is, $p \leq 0.05$, then we would have to look at the pairwise comparisons (A vs. B, A vs. C, or B vs. C) to determine where the difference occurred. This is called a multiple comparison procedure and requires an adjustment to the p-value.

Since significance at the 0.05 level in the ANOVA was not attained, we would not bother with any pairwise comparisons, since the simultaneous test of the equality of the means would show no difference. However, for the sake of completeness, and since we will apply this multiple comparison procedure in future situations, we explain the procedure here in the following section.

2.3.4.1 Multiple Comparison Procedure

Tukey–Kramer Test A test developed by Tukey (1953) and later refined by Kramer (1956) is used extensively to make pairwise comparisons among means. The test is often referred to as the Tukey–Kramer test or the HSD (honestly significant difference) test. It makes use of a single value against which the pairwise mean comparisons are made at the appropriate α level of significance. The value HSD is given by the formula

$$\text{HSD} = q_{\alpha,k,N-k} \sqrt{\frac{\text{MSE}}{2} \left(\frac{1}{n_i} + \frac{1}{n_j} \right)}. \tag{2.33}$$

In general, α is obviously the level of significance, which in most cases is 0.05, k is the number of means in the experiment and N is the total sample size and $N - k$ is called the error degrees of freedom or error df. The quantity, q, is obtained from published statistical tables for the values of α, k, and $N - k$ (Pearson and Hartley 1958). The value, MSE, is the mean-squared error or the within groups means square (MS) as seen, for example, in Table 2.10B. The values n_i and n_j are the sample sizes of the individual groups i and j where in our aforementioned case i and j can take on the labels of A, B, or C, since we have three groups as A, B, and C. The procedure for determining if two group means differ statistically from each other is very simple. One compares the actual computed absolute mean differences to the computed HSD for the two groups. If the absolute difference is greater than HSD, then the two means are considered statistically different at the appropriate level of α. If the absolute difference is not greater than HSD, then the two group means are not statistically different. Keep in mind that this procedure is only done if the overall p-value for the ANOVA is less than or equal to 0.05, which is not the case for Table 2.10. However, for demonstration purposes, we will show the calculations using these means.

Example Note in Table 2.10A, the mean HDL for group A is 47.72 and for Group B it is 46.036. Thus, the absolute difference is $|47.72 - 46.036| = 1.684$. The MSE from Table 2.10B is 81.955. Also, $N = 95$, $k = 3$, $N-k = 92$, $n_A = 37$ and $n_B = 39$ (Pearson and Hartley 1958). We have the value of $q_{\alpha,\,k,N-k} = q_{0.05,\,3,\,60} = 3.40$ and $q_{0.05,\,3,\,120} = 3.36$. So obviously, the value for our example or $q_{0.05,\,3,\,92}$ lies somewhere between 3.36 and 3.40. To get the exact value, we interpolate using the following technique. For $\alpha = 0.05$ and the $N - k$ values we have

$N - k$	q
60	3.4
92	X
120	3.36

We must solve (interpolate) for the value of X using the formula

$$\frac{92 - 60}{120 - 60} = \frac{X - 3.40}{3.36 - 3.40}. \tag{2.34}$$

Note that we have the same formulation for the numerators and denominators on both sides of the equation and thus solve for X, which is our value for q for $N - k$ or error df $= 92$. Solving for X yields the value $q = 3.3787$, which makes sense since it is between the values of 3.36 and 3.40.

Thus for means A and B the HSD is

$$\text{HSD} = 3.3787 \sqrt{\frac{81.955}{2}\left(\frac{1}{37} + \frac{1}{39}\right)}$$

$$= 4.964. \tag{2.35}$$

Note that $|47.72 - 46.036| = 1.684 < \text{HSD} = 4.964$. Thus, these two means are not statistically different from each other. We do the same for A versus C and B versus C in Table 2.11.

Although the comparisons would not have been performed ordinarily, we see that they are consistent with the overall nonrejection of the null hypothesis of the equality of the three means. We will revisit interpolation in another context in a later chapter, and we will revisit the Tukey–Kramer test of multiple comparisons in Chapter 7.

TABLE 2.11 Absolute Mean Differences and HSD Pairwise Comparisons

Comparison	Absolute Mean Difference	HSD	Means Statistically Different?
A versus B	1.684	4.964	No
A versus C	2.891	6.104	No
B versus C	1.207	6.051	No

Bonferroni Comparisons One of the most commonly used multiple comparison procedures is called the Bonferroni comparison procedure. The test statistic for the Bonferroni comparison procedure is given by the formula

$$t = \frac{\bar{x}_i - \bar{x}_j}{\sqrt{MSE\left(\frac{1}{n_i} + \frac{1}{n_j}\right)}} \tag{2.36}$$

For a two-sided level α test, $\alpha^* = \frac{\alpha}{\binom{k}{2}}$, where k is the number of groups. By doing this, we want to ensure the overall probability of making a false claim is maintained at α for all possible pairwise comparisons. For k groups, $\binom{k}{2}$ possible pairwise comparisons can be made. The expression $\binom{k}{2}$ is the combination of a pair to choose from k groups at a time. It represents the number of ways two groups can be selected from a total of k groups without regard to order. $\binom{k}{2} = \frac{k!}{2!(k-2)!}$, where $k!$ or k factorial is calculated as $k * (k-1) \ldots 2 * (1)$. For total of three groups, $\binom{3}{2} = \frac{3!}{2!(3-2)!} = \frac{3*2*1}{2*1(1)} = 3$ ways to choose a pair to compare. For example, for three group comparisons with $\alpha = 0.05$, $\alpha^* = \frac{0.05}{\binom{3}{2}} = \frac{0.05}{3} = 0.0167$. Therefore, conduct t-tests between each pair of groups using the new α level of $0.0167/2 = 0.00835$ of significance for a two-sided t-test.

Example Note in Table 2.10, the mean HDL for group A is 47.72 and for group B it is 46.036. Thus, the absolute difference is $|47.72 - 46.036| = 1.684$. The MSE from Table 2.10B is 81.955. Also, $k = 3$, $n_A = 37$ and $n_B = 39$. We have the t-statistic

$$t = \frac{47.72 - 46.036}{\sqrt{81.955\left(\frac{1}{37} + \frac{1}{39}\right)}} = \frac{1.684}{2.078} = 0.811.$$

Note that $|47.72 - 46.036| = 1.684 < 2.449 =$ the t-critical value of 74 df ($n_A + n_B - 2 = 37 + 39 - 2 = 74$). Therefore, these two means are not statistically different from each other. We do the same for A versus C and B versus C in Table 2.12.

TABLE 2.12 Absolute Mean Differences and Pairwise Comparisons Using Bonferroni Comparisons

Comparison	t-Statistic	t-Critical Value	Degrees of Freedom	Means Statistically Different?
A versus B	0.811	2.449	74	No
A versus C	1.131	2.470	54	No
B versus C	0.477	2.467	56	No

We see that they are consistent with the overall nonrejection of the null hypothesis of the equality of the three means. Furthermore, the results are consistent with the Tukey–Kramer test of multiple comparisons.

2.3.5 Nonparametric Tests for Skewed Data

When the data set is not normally distributed, we cannot use typical tests such as t-tests or ANOVA, as they are predicated upon the normal distribution. Figure 2.13 shows that the data is right skewed (not normal) for a dilution experiment resulting in values of potassium for dilution and mock groups, more so in the mock group. The summary data and the statistics are in Table 2.13. Thus, we use a nonparametric (NP) test to compare the two distributions, not assuming that the data have any particular parametric distribution such as normality. The focus of a nonparametric test of hypothesis is the comparison of the likelihood that the data from different groups came from the same distribution or follow a similar distribution. Nonparametric (NP) tests do not necessarily focus the inference on a parameter such as the mean which we have been doing earlier under the assumption of normality. These NP tests follow the same procedures as above for hypothesis testing. That is to say, follow the same four steps in the hypothesis testing approach that we outlined in equation (2.12).

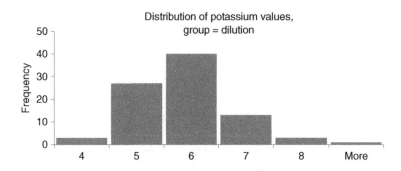

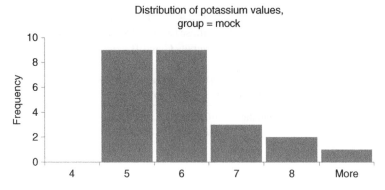

Figure 2.13 Skewed Potassium Values by Group

TABLE 2.13 Descriptive Statistics and Generic Output from a Wilcoxon (Mann–Whitney) Test for the Skewed Potassium Values

Group	Dilution	Mock
Mean	5.378161	5.5875
Standard error	0.092868	0.222902
Median	5.3	5.15
Standard deviation	0.866217	1.091992
Minimum	3.6	4.3
Maximum	8.2	8.2
N	87	24
Wilcoxon (Mann–Whitney)	—	—
Two-sided p-value	0.7963	—

Thus in our present example, we wish to compare the two distributions (dilution vs. mock) to determine if they are the same.

The null hypothesis is

$$H_0: \text{ Distributions are the same versus } H_A: \text{ Distributions are not the same.} \quad (2.37)$$

Here, we are not comparing parameters such as means or variances, but the entire distributions.

A nonparametric test still yields a p-value to test the hypotheses. NP results can be easily calculated using any statistical software. They are generally too computationally heavy to do by hand. In this case, we use what is called the Wilcoxon test, found in most statistical references (Rosner 2010). The p-value is 0.7963, which as we know from our four-step procedure that $p > 0.05$. Thus, we do not reject the null hypothesis that the distributions are the same and conclude the potassium levels for both the mock and dilution groups are the same. Again note the skewness on the histogram plots, especially the mock group. Therefore, it is appropriate to use the Wilcoxon test. NP procedures are often used in the laboratory hypothesis testing framework.

REFERENCES

Joanes DN and Gill CA. (1998). Comparing measures of sample skewness and kurtosis. Journal of the Royal Statistical Society (Series D): The Statistician 47 (1): 183–189.

Kramer CY. (1956). Extension of multiple range tests to group means with unequal sample sizes. Biometrics 63: 307–310.

Owen WJ and Siakaluk PD. (2011). A demonstration of the analysis of variance using physical movement and space. Teaching of Psychology 38(3): 151–154.

Pearson ES and Hartley HO. (1958). Biometrika Tables for Statisticians, 2nd ed., Cambridge University Press, New York Vol. 1, Table 29.

Rosner B. (2010). Fundamentals of Biostatistics, 7th ed., Thomson-Brooks/Cole, Belmont, CA. (Duxbury Press.)

Tukey JW. (1953). The Problem of Multiple Comparisons. Mimeographed Monograph. Princeton University. cited in Hochberg Y and Tamhane AC. Multiple Comparison Procedures. John Wiley and Sons, New York. 1987.

3

METHOD VALIDATION

3.1 INTRODUCTION

This chapter is actually an introduction to validation procedures in general with a discussion of accuracy from two perspectives. One approach using linearity compares the results of a new method versus a standard method. The second methodology discusses spiking the sample matrix with a known concentration of reference material and extracting and measuring the analyte from the matrix and comparing it with a reference sample when added to pure solvent. There is a thorough discussion of understanding the linear model and testing of the slope and intercept and residual analysis. Mandel sensitivity is discussed when both methods are considered random. The chapter then moves on to selectivity (specificity) and sensitivity. Relative potency using the linear model is also presented.

According to the United States Food and Drug Administration (FDA) Guidance for Industry on Bioanalytical Method Validation (2001), "Selective and sensitive analytical methods for the quantitative evaluation of drugs and their metabolites (analytes) are critical for the successful conduct of preclinical and/or biopharmaceutics and clinical pharmacology studies. Bioanalytical method validation includes all of the procedures that demonstrate that a particular method used for quantitative measurement of analytes in a given biological matrix, such as blood, plasma, serum, or urine, is reliable and reproducible for the intended use. The fundamental parameters for this validation include (1) accuracy, (2) precision, (3) selectivity, (4) sensitivity, (5)

Introduction to Statistical Analysis of Laboratory Data, First Edition.
Alfred A. Bartolucci, Karan P. Singh, and Sejong Bae.
© 2016 John Wiley & Sons, Inc. Published 2016 by John Wiley & Sons, Inc.

reproducibility, and (6) stability. Validation involves documenting, through the use of specific laboratory investigations, that the performance characteristics of the method are suitable and reliable for the intended analytical applications. The acceptability of analytical data corresponds directly to the criteria used to validate the method." Added to the six parameters listed for the food processing industry, the Environmental Protection Agency (EPA) and manufacturing industry would add robustness, ruggedness, limits of detection, and limits of quantitation, among others. We discuss all of these terms in the chapters that follow.

Let's start out with a few definitions and ground rules before we proceed.

An analyte is the substance or chemical constituent that is undergoing analysis. It is the substance being measured in an analytical procedure. An example of an analytical procedure is titration. For instance, in an immunoassay, the analyte may be the ligand or the binder, while in blood glucose testing, the analyte is glucose. In medicine, "analyte" often refers to the type of test being run on a patient. This is because the test is usually measuring a chemical constituent in the body.

Method validation is the process for establishing that performance characteristics of the analytical method are suitable for the intended application. The validity of an analytical method can only be verified by laboratory studies. All variables of the method should be considered, including sampling procedure, sample preparation, detection, and data evaluation using the same matrix as that of the intended sample. Matrix effect (in analytical chemistry) is the combined effect of all components of the sample (e.g., soil made up of clay, sand, and chalk), that is, variables other than the analyte, on the measurement of the quantity. If a specific component can be identified as causing an effect, then this is referred to as interference. Liquid chromatography/mass spectrometry (LC/MS) with electrospray ionization is a highly specific and sensitive analytical technique that has become the industry standard for quantifying drugs, metabolites, and endogenous compounds in biological matrices (e.g., plasma). The technique is widely used because of its ability to accurately quantitate analytes of interest with minimal sample clean-up and rapid LC separation.

Despite these positives, LC/MS utilization will at times encounter problems, some of which are caused by matrix effects. As Hall et al. (2012) note in drug discovery, some common matrices typically encountered by scientists and analytical chemists during bioanalysis include blood, plasma, urine, bile, feces, and tissue samples. It is known that these complex matrices have a number of common components, but not all are known and the levels may vary among individuals. An example in clinical trials is that plasma samples obtained from different subjects enrolled in the study may contain different levels of endogenous components based on their genetics and/or disease state, as well as different drugs or interventions used to manage their disease. One such intervention could be concomitant medications taken by an individual, which may influence outcome of a targeted drug. Thus, each patient's plasma may have its own particular set of matrix components and therefore is viewed as being unique to the patient. Thus, in the LC/MS method, the degree of enhancement or suppression of ionization of an analyte by a given matrix component can be dependent on the physicochemical properties of the analyte. Matrix effects can be a complicated and time-consuming challenge for the analytical chemist charged with developing robust, reproducible, and accurate analytical methods. There are a number of diverse sample and system conditions that lead to matrix effects and an equally diverse set of potential options to remedy them. We should be aware of these possibilities when we discuss our statistical treatment of the data. We will in fact in later chapters discuss statistical ways of quantifying factors that affect the variation in an analyte response. This is not to be confused with the complex chemical analytic methodology of extracting and/or detecting matrix effects.

3.2 ACCURACY

Accuracy of an analytical method is the degree of agreement of test results generated by the method to the true value.

This can be done in either of the following two ways.

Method 1: Compare the results of the method with results of an established reference method.

or

Method 2: Spike the sample matrix with a known concentration of reference material, extract and measure the analyte from the matrix, and compare its response with the response of the reference material when added to pure solvent.

We'll discuss each method in the following subsections.

3.2.1 Method 1

Compare the results of the new method with results of an established reference method.

We use an example from data generated when comparing two methods of performing chemical assays. Consider the Abbott architect c8000™ clinical chemistry system compared to its own standard AEROSET system. Using the AEROSET system as the standard that has a positive track record, how accurate is the c8000 with respect to its analysis of chemical assays? In the way of further introduction as per Vannest et al. (2002), the ARCHITECT c8000, a mid-volume chemistry analyzer, was designed with the same optics and fluidics characteristics of the Abbott AEROSET, a high-volume clinical chemistry analyzer. The c8000 can be utilized as a stand-alone analyzer, or it can be integrated with the Abbott ARCHITECT i2000SR Chemiluminescent Immunoassay analyzer issued to perform a large number of immunoassays and traditional chemistry applications. The c8000 utilizes the same reagents and assay parameters as the AEROSET.

The purpose of their study was to evaluate commutability of traditional photometric assay results between the c8000 and the AEROSET and between c8000 instruments. Precision, method comparison versus AEROSET, limit of quantitation and linearity studies were conducted on three c8000 instruments to evaluate

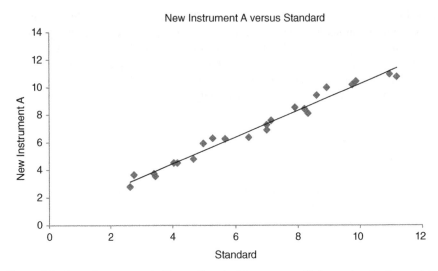

Figure 3.1 New Instrument A Versus Standard Instrument (Reference) with Respect to Measuring a Particular Analyte

consistency of performance between c8000 and AEROSET. Here we consider the method comparison study. One compares the results of the new method with results of an established reference method. We discuss a similar method in general with our own example. In Figure 3.1, the clinical laboratory results of a measured analyte (e.g., normal and abnormal BUN-Blood Urea Nitrogen mmol/L) from a new versus standard instrument are plotted. This is a simple linear comparison, and one merely sees the homogeneous nature of the readings of the two methods versus each other. The statistical strength of the relationship will be examined and explained. However, before we do, we discuss the statistical concept of linearity.

3.2.1.1 Linearity We will discuss linearity based on the previous accuracy example.

To construct a valid linear relationship, the conditions of linearity are as follows:

– There should be six or more calibration standards or points over the concentration range.
– The calibration standards should be evenly spaced over the concentration range of interest.
– The range should be 0–150% or 50–150% of concentration likely to be encountered, depending on which is more suitable.
– The calibration standards should be run at least in duplicate, preferably in triplicate or more in random order.

For statistical analysis considerations, refer to Figure 3.1. We ask the following:

- Is the calibration function
  - Linear (check the fit via computer output and residuals – we will explain these)?
  - Does the line pass through the origin (the 0,0 coordinate on the graph)?

From the analytical chemistry perspective, can it be shown or assumed that

- The result is unaffected by the matrix of the test material?

The following is an example examining linearity statistically.

For the sake of demonstration, we assume we have the data in Table 3.1, which are 18 response readings from two instruments (standard vs. test) recorded over a valid range of values from one to six. They can be considered as paired laboratory readings (one from the standard instrument and one from the test instrument) taken on a blood sample from 18 subjects. The plot of the data is shown in Figure 3.2. The plot called the least squares line was generated from the linear regression option in EXCEL.

The equation of the line is

$$Y = -0.0308 + 1.0088X, \qquad (3.1)$$

TABLE 3.1 Data for Standard Versus Test Methodology

Subjects	Standard	Test	Predicted Test	Residuals
1	1.5	1.6	1.48	0.12
2	1.5	1.4	1.48	−0.08
3	1.7	1.6	1.68	−0.08
4	2	2	1.99	0.01
5	2.3	2.31	2.29	0.02
6	2.7	2.7	2.69	0.01
7	3.6	3.6	3.60	−0.00
8	3.5	3.6	3.50	0.10
9	3.4	3.3	3.40	−0.10
10	3.9	4	3.90	0.10
11	3.8	3.7	3.80	−0.10
12	3.96	3.9	3.96	−0.06
13	4.3	4.3	4.31	−0.01
14	4.6	4.7	4.61	0.09
15	4.5	4.6	4.51	0.09
16	5.7	5.8	5.72	0.08
17	5.71	5.7	5.73	−0.03
18	5.63	5.5	5.65	−0.15

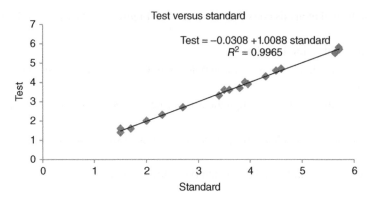

Figure 3.2 Test vs. Standard Results (Data from Table 3.1)

where $Y = \text{TEST}$ and $X = \text{STANDARD}$. Often a more common format of the equation is written as $Y = -0.0308 + 1.0088X$ or $\text{TEST} = -0.0308 + 1.0088\,\text{STANDARD}$. We discuss linearity in general in the following: the linear relationship between a variable Y, often called the response, and the variable X, often called the predictor or independent variable is

$$Y = \beta_0 + \beta_1 X. \tag{3.2}$$

β_0 is called the intercept. This is the value where the line crosses the Y (vertical) axis. β_1 is called the slope. The variable, X, is multiplied by β_1. β_0 and β_1 are called parameters that we estimate from the data. Note that for every unit change in X the value of Y increases by β_1. For example, if we increase X by 1, then the equation becomes

$$Y = \beta_0 + \beta_1(X + 1) = Y = \beta_0 + \beta_1 + \beta_1 X, \tag{3.3}$$

and obviously Y has increased by a value of β_1. Figure 3.3 demonstrates this concept. Clearly, as X increases by 1 unit on the horizontal axis, Y increases by β_1 units on the vertical axis. In our aforementioned model, the TEST versus the STANDARD has the form $\text{TEST} = -0.0308 + 1.0088 * \text{STANDARD}$.

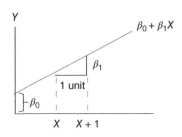

Figure 3.3 Increase in Y per One Unit Increase in X

The estimate of β_0 or the intercept is -0.0308 and the estimate of β_1 or the slope is 1.0088. Thus, for every unit change in the Standard, the test value increases by a value of 1.0088.

One other statistic generated in Figure 3.2 is called the R^2 (R-square) statistic. Note that $R^2 = 0.9965$, which is considered strong. The range of the R^2 is between 0 and 1.0. One interprets the R^2 as the amount of variation in the Y variable explained by the X. In our case, the amount of variation or change in the Test method explained by the Standard is about 0.9965 or, expressed as a percent, is about 99.65%. Obviously, the closer the R^2 to 1.0 means a stronger association between the two methods and an R^2 closer to 0 is a weak association between the two methods. We could thus say that it appears that there is a strong association between the two methods or instruments or the TEST is certainly well validated by a linear function of the Standard. We are not yet discussing "agreement" per se, which we will. A further statistical interpretation in terms of hypothesis testing will be discussed later. Also, keep in mind that R^2 is not a measure of the appropriateness of the linear model. It measures the strength of the relationship of the variables to each other, but that relationship need not be linear. It could be a curvilinear relationship. Or, stated differently, once you decide on the appropriateness of the model, the closer its R^2 value is to 1, the greater the ability of that model to predict a trend.

Let's review the major assumptions of the linear model:

1. The response Y (new or Test) and the predictor variable (standard) are linearly related.
   Note here we often call X the predictor variable. It need not be a predictor. We are really interested in the strength of the association between X and Y. We'll discuss the meaning of the strength of association later.
2. The data points (Ys) are independent and normally distributed as per our discussions of Chapter 2.
3. The variance of the points about the line is constant. In other words, you can visually look at the line and the scatter of the points about the line, and then ask yourself, "Does a linear fit make sense?"

We'll discuss this as well.

Often, we should consult with a statistician, and he or she can easily test if the appropriate assumptions are met to allow us to say we have a true linear relationship between the two variables. Formal statistical testing for linearity or lack of fit is beyond the scope of this book. However, we'll provide useful tools so that one can make a pretty good assessment of the appropriateness of a linear fit. Let's consider point number (3) of the above list and discuss the residual plot for determining linearity.

The residuals are the distance between the actual observed Y value and the predicted Y value from the equation. For example, let's take the data in Table 3.1 and apply it to equation (3.2). Recall the estimate of the intercept or $\beta_0 = -0.0308$ and the estimate of the slope or $\beta_1 = 1.0088$. Thus, the estimate of the predicted equation is $Y_{\text{Pred}} = -0.0308 + 1.0088X$ for a value of X. Let's take the first observation in

Table 3.1 as an example. Here X or standard value is 1.5. Therefore, the predicted value of Y or the Test is

$$Y_{\text{Pred}} = -0.0308 + 1.0088(1.5) = 1.48. \tag{3.4}$$

Note the actual observed value for Y or the Test $= 1.6$.

The residual for the ith observation that we label as e_i for every point is defined as the difference of the observed value, Y, and the predicted value of Y. That is,

$$e_i = Y_i \text{ observed } - Y_i \text{ predicted}, \ i = 1, \ \dots \ , n.$$

The assumptions on the residuals are as follows: e_i are randomly scattered about the predicted line with constant variation. Their mean $= 0$.

The e_i are normally distributed and are thus randomly scattered about their mean, 0. Therefore, for the first observation, or $i = 1$, the residual value is

$$e_1 = 1.60 - 1.48 = 0.12.$$

Let's take another observation, say observation 17. Here $Y = 5.7$ and $X = 5.71$. Thus, the predicted value for Y is $Y_{\text{Pred}} = -0.0308 + 1.0088(5.71) = 5.73$ and the residual is $5.7 - 5.73 = -0.03$. Note in Table 3.1, the last two columns give the predicted values of Y and the residual for all the observations. One can see that there are negative and positive residuals giving a random scatter about zero, which is the residual mean. Figure 3.4 shows by example in a proper linear situation how the residuals will scatter about the predicted line. The predicted line is actually the mean residual line. Revisiting Figure 3.2, one can see reasonable scatter of the points about the line. One often examines the appropriateness of a linear model by plotting the predicted values on the horizontal axis and the residuals on the vertical axis, and if there is good constant scatter about the residual mean line of zero, then the constancy of variance or scatter is satisfied. Figure 3.5, the residual plot for this data, confirms the model in this case.

If there were perfect agreement between the Test and the Standard in some cases, one would expect the line relating the two methodologies to have a slope very close to one and the intercept would be near zero or the line passes through the origin.

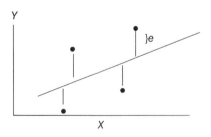

Figure 3.4   What Are the Residuals?

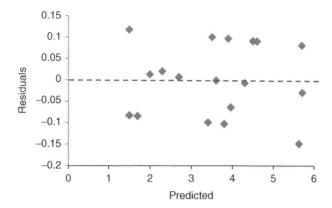

Figure 3.5   Residuals Versus Predicted Values (Data from Table 3.1)

One can think of the line being at about a 45° angle or for every unit change in X, the Y value changes by a unit or the slope $= 1$. This, of course, is not necessarily realistic. However, one can then conduct a test that the values of the parameters not be significantly different from a hypothesized value, say 0 for the intercept and 1 for the slope. Most statistical packages will provide a test for both parameters equal to 0. We test for the slope $= 1$ as well. We refer to our hypothesis testing skills outlined in Chapter 2 and conduct such a test. Refer to Table 3.2, which is the regression results for our method validation example. Recall we want to know: Is the intercept of our line $= 0$ or does the line pass through the origin? Is the slope $= 1$? The t-statistic or t-ratio for testing the intercept $= 0$ or for testing the hypotheses at the 0.05 alpha level in row 1 of the table is

$$t = \frac{(-0.0308 - 0)}{(0.0571)} = -0.54, \tag{3.5}$$

where -0.0308 is the estimate of the intercept, β_0, from the data, 0 is the hypothesized value of the intercept and 0.0571 is the standard error of the estimate. Note the p-value here is 0.596, or is greater than 0.05, or we do not reject the null hypothesis and conclude the intercept is not statistically different from 0.

TABLE 3.2   Parameter Values for the Validation Example

Row	Parameter	Estimate	Std. Error	Hypotheses	t-Ratio	p-Value
1	Intercept (β_0)	-0.0308	0.0571	$H_0: \beta_0 = 0$ $H_1: \beta_0 \neq 0$	-0.539	0.596
2	Slope (β_1)	1.0088	0.0149	$H_0: \beta_1 = 0$ $H_1: \beta_1 \neq 0$	67.70	0.001
3	Slope (β_1)	1.0088	0.0149	$H_0: \beta_1 = 1$ $H_1: \beta_1 \neq 1$	0.591	0.718

The t-statistic or t-ratio for testing the slope $= 0$ or for testing the hypotheses at the 0.05 alpha level in row 2 of the table is

$$t = \frac{(1.0088 - 0)}{(0.0149)} = 67.70, \tag{3.6}$$

where 1.0088 is the estimate of the slope, β_1, from the data, 0 is the hypothesized value of the slope and 0.0149 is the standard error of the estimate. Note the p-value here is 0.001 or less than 0.05, or we do reject the null hypothesis, $H_0 : \beta_1 = 0$, and conclude the slope is statistically different from 0. This is confirmation of a statistically significant association between the Y variable and the X variable. If we did not reject this null hypothesis, then we would be concerned that the association between the two variables would be weak or nonexistent, and this translates in our case of validation that the method agreement is very weak. Thus, one may not wish to pursue the issue further.

However, since we do reject that the slope $= 0$, let's take it one step further and test that the slope $= 1$. This is comparable to testing that for every unit change in X we have a unit change in Y or very good agreement between the two methods. The t-statistic or t-ratio for testing the slope $= 1$ or for testing the hypotheses at the 0.05 alpha level in row 3 of the Table 3.2 is

$$t = \frac{(1.0088 - 1)}{(0.0149)} = 0.591, \tag{3.7}$$

where, again, 1.0088 is the estimate of the slope, β_1, from the data, 1 is the hypothesized value of the slope and 0.0149 is the standard error of the estimate. Note the p-value here is 0.718, or certainly greater than 0.05, or we do not reject the null hypothesis and conclude the slope is not statistically different from 1.

The method agreement is significant. Our method comparison and relative bias analysis of results between the standard and the test instruments demonstrated commutability of results between the two platforms and reproducibility of results between the two instruments. We will discuss the bias concept later in Chapter 7.

Let's consider another example from the food processing industry. Curiale et al. (1991) considered method comparison for reproducibility of total coliforms and *Escherichia coli* (*E. coli*) in foods. The three methods considered were as follows: PCC or petricoliform count plates, PEC or petrifilm count plates, and MPN or most probable number methods. PEC and PCC are called the dry film methods. In one such analysis, they examined the total coliform counts for turkey samples by the dry film plate count methods and the MPN method. Food was contaminated (inoculated) with low, medium, and high levels of bacteria. We consider the medium-level analysis here. Figure 3.6 demonstrates the linear relationship between the two methods of PCC, which we label as the standard (X) and PEC as the TEST (Y). The units are in log total coliform counts/gram. One can see from the results shown in Table 3.3 and Figure 3.6 that the two methods are comparable. The equation of the line is

$$PEC = 0.119 + 0.948\,PCC \tag{3.8}$$

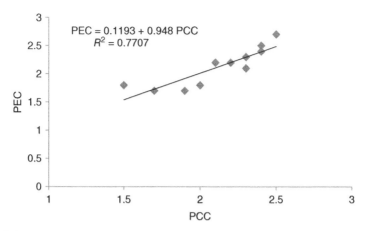

Figure 3.6 Method Comparison PEC Versus PCC *Escherichia coli* Data from Curiale et al. (1991)

TABLE 3.3   Parameter Values for the Validation Example of *Escherichia coli* (PCC vs. PEC)

Row	Parameter	Estimate	Std. Error	Hypotheses	t-Ratio	p-Value
1	Intercept (β_0)	0.119	0.368	$H_0 : \beta_0 = 0$ $H_1 : \beta_0 \neq 0$	0.323	0.754
2	Slope (β_1)	0.948	0.172	$H_0 : \beta_1 = 0$ $H_1 : \beta_1 \neq 0$	5.912	0.001
3	Slope (β_1)	0.948	0.172	$H_0 : \beta_1 = 1$ $H_1 : \beta_1 \neq 1$	−0.302	0.161

Source: Data from Curiale et al. (1991).

with an $R^2 = 0.7707$. Also testing our hypotheses at the 0.05 level, we have an intercept not statistically significantly different from 0 (p-value $= 0.754$) and the slope is significantly different from 0 (p-value $= 0.001$) and not statistically different from 1 (p-value $= 0.161$). We have commutability of the two methods. The residual plot (not shown) confirms that the model is adequate.

Now let's consider the MPN versus the PEC. Figure 3.7 demonstrates the linear relationship between the two methods of PEC, which we label as the standard (X) and MPN as the TEST (Y). The units are in log total coliform counts/gram. One can see from the results in Table 3.4 and Figure 3.7 that the two methods are comparable. However, this is somewhat of a weaker association. The equation of the line is MPN $= 0.883 + 0.692\,$PEC with an $R^2 = 0.384$. Also, testing our hypotheses at the 0.05 level, we have an intercept not statistically significantly different from 0 (p-value $= 0.193$) and the slope is significantly different from 0 (p-value $= 0.042$) and not statistically different from 1 (p-value $= 0.341$). The residual plot (not shown)

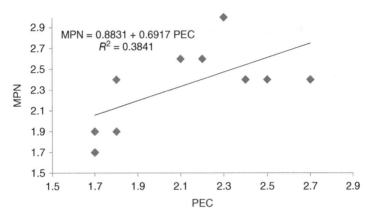

Figure 3.7   Method Comparison MPN Versus PEC *E. coli* Data from Curiale et al. (1991)

TABLE 3.4   Parameter Values for the Validation Example of *E. coli* (MPN vs. PEC)

Row	Parameter	Estimate	Standard Error	Hypotheses	t-Ratio	p-Value
1	Intercept (β_0)	0.883	0.628	$H_0 : \beta_0 = 0$ $H_1 : \beta_0 \neq 0$	1.406	0.193
2	Slope (β_1)	0.691	0.292	$H_0 : \beta_1 = 0$ $H_1 : \beta_1 \neq 0$	2.367	0.042
3	Slope (β_1)	0.948	0.292	$H_0 : \beta_1 = 1$ $H_1 : \beta_1 \neq 1$	−0.178	0.341

Data from Curiale et al. (1991).

confirms that the model is adequate. We have commutability of the two methods. However, with an $R^2 = 0.384$, one would question the strength of the commutability. Although these methods demonstrate the technique, a preferable relationship for comparability of methods is seen in the data in Table 3.1, where the R^2 is high, about 0.99, the slope is very close to 1 and the intercept is close to 0 as seen in Table 3.2.

3.2.1.2   Mandel Sensitivity   This is another way of method comparison by Mandel and Stiehler (1954), which is an old method, but still considered practical. It involves both the standard deviation of each method and the rate of change in readings by each method as the underlying quantity being measured changes. It actually tells you how many replicated readings of one method on average will give the same precision as the other method.

Let sd(A) be the standard deviation of method A (Standard) and sd(B) be the standard deviation of method B (Test). Here, we assume both methods A and B are random quantities. What do we mean by this last statement? In the linear regression procedure we discussed in the last section, the assumption is that the response, Y, is a random quantity measured with some error, and the predictor or X variable is a

TABLE 3.5 Mandel Sensitivity Data

Observation	Method A	Method B
1	3.5	2.5
2	2	2.5
3	2.6	3
4	2.6	2.7
5	3.2	3.5
6	2.5	2.5
7	3.9	4
8	4.2	4.1
9	4.3	4.4
10	5.1	5.4

nonrandom fixed value. Thus, for each X there is a random response, Y. The Mandel sensitivity allows both X and Y to be random. Without going into the theoretical derivation of the quantities needed for Mandel sensitivity, we present the basic tools for conducting this analysis and interpret what they are telling us. We refer to the data of Table 3.5.

Let K be slope of the line relating A to B. The advantage of the Mandel approach is that it can be used for a general relationship of A to B. In other words, the relationship need not be linear. One computes the sensitivity of one method to another in a region of interest. That is to say, K is the slope of the curve in the region where a comparison of interest is made. It may not be over the entire region of the curve. It may be over a particular range of the data. For our purpose, we have a linear relationship of A to B in Table 3.5, and we will consider the entire range of the data. The sensitivity measure, SM, is defined as

$$SM = |K| \times \left[\frac{sd(A)}{sd(B)} \right]. \tag{3.9}$$

The slope for random effects is $K = 0.898$. This is not the usual slope from a linear regression assuming A is fixed and B is random. Both measures are random. The slope for the linear relationship of A to B for nonrandom A is 0.921. One can see they are close but not exactly the same. One uses a more sophisticated program such as PROC MIXED in SAS (version 9.3) to derive this random slope value. Continuing with our calculations, $sd(B) = 0.994$ and $sd(A) = 0.982$. Note A has slightly more precision. Thus,

$$SM = 0.898 \times \left[\frac{sd(A)}{sd(B)} \right] = 0.887 \tag{3.10}$$

If SM exceeds unity, this tells us that B is superior to A in precision over the range of the data, which obviously is not. We square SM to get $(0.887)^2 = 0.787$.

If we calculate $1/(0.787) = 1.27$, then we need 1.27 measurements of B to get the same precision as A. In other words, for about every four measurements of A, we need $1.27 \times 4 = 5.08$ or about five measurements of B to attain the same precision. According to the Mandel approach, one can also put a lower 95% confidence limit on the sensitivity measure, SM. This is simply

$$\text{SM} \times \frac{1}{\sqrt{F}}, \tag{3.11}$$

where F is the F-distribution critical value on $(9,9)$ degrees of freedom (df). See Chapter 2 for a discussion of the F-distribution. We have 9 df since we have 10 observations on A and B and $10 - 1 = 9$ df for each method. The critical value is, thus, 3.18, and we have

$$\text{SM} \times \frac{1}{\sqrt{F}} = 0.887 \times \frac{1}{\sqrt{3.18}} = 0.497. \tag{3.12}$$

Doing the same calculations as above, the worst-case scenario is that we would require about four observations of B for every A to attain the same precision. This lower confidence limit is not often calculated, but it is interesting to note.

3.2.2 Method 2

The previous sections showed accuracy as a linear relationship. We are going to stay in the linearity mode and consider a second definition of an accuracy method, that is, spike the sample matrix with a known concentration of reference material, extract and measure the analyte from the matrix, and compare its response with the response of the reference material when added to pure solvent.

We want to examine the accuracy of a New Gene Measuring System (NGS) instrument in measuring mRNA. Some definitions are in order.

Definitions
- mRNA = messenger RNA, which is the copy of the information carried by a gene on the DNA measured by three cytokine genes and a "housekeeping" gene in this setting
- Buffer stabilizes the pH (acidity or alkalinity) of a liquid
- DAP = death associated protein involved in TNF α (tumor necrosis factor alpha)
- attomole = 1×10^{-18} mol = 600,000 molecules
- RLU = relative luminescence.

One goes through some standard procedures in order to measure accuracy. A dilution series of bacterial control RNA (DAP) ranging from 5 to 40 attomoles is added to buffer to get a standard curve. From the generation of the standard curve (reference), two known concentrations of the DAP RNA $(8, 20)$ attomoles are added to the buffer (spiked) and measured by the NGS.

Table 3.6 is the data for this example and Figure 3.8 shows the results. The solid line shows the reference and the dashed line shows the NGS measured recovery of mRNA from the spiked sample.

TABLE 3.6   Data for New Gene Expression

Observation	Type	Attomoles	RLU
1	Standard	5	10
2	Standard	7	30
3	Standard	10	90
4	Standard	20	180
5	Standard	30	270
6	Standard	40	440
7	Spiked	7	35
8	Spiked	10	95
9	Spiked	20	190
10	Spiked	40	445

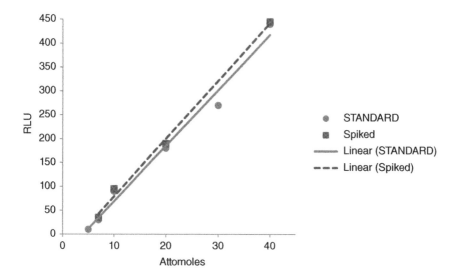

Figure 3.8   Standard Versus Spiked Sample

Table 3.7 and Figure 3.9 show the results of the aforementioned procedure. The data in Table 3.7 refers to the results in Figure 3.9 of how the NGS calculated values are derived. In summary, one computes the regression line curve for the (buffer–solid line) reference data and solves for the RLU for RNA DAP of 8 and 20. One then derives the equation using the NGS system (spiked–dashed line) and uses the RLU derived for 8 and 20 from the reference, then inserts that RLU value into the spiked equation and solves for the mRNA attomole value.

The percent recoveries are $7.3/8.0 = 91.25$ and $18.8/20 = 94.0$.

Here is the calculation:

The reference equation for the Standard data in Table 3.6 is

$$RLU = -46.786 + 11.614 \times \text{attomoles}. \qquad (3.13)$$

TABLE 3.7   Data for New Gene Expression

RLU	Actual (Solid)	NGS Calculated (Dashed)	% Recovery
46.1	8	7.3	91.25
185.2	20	18.8	94.0

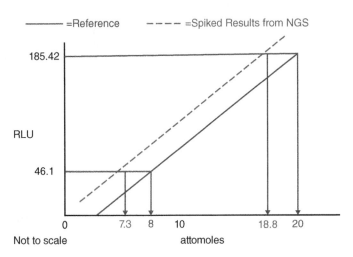

Figure 3.9   Accuracy Gene Expression

For attomoles $= 8.0$ in equation (3.13) RLU $= 46.1$ (first row, first column of Table 3.7).

The equation for the NGS (Spiked data in Table 3.6) is

$$RLU = -41.706 + 12.102 \times \text{attomoles} \qquad (3.14)$$

Let RLU $= 46.1$ in equation (3.14) and solve for attomoles, then attomoles $= 7.3$ (first row, third column in Table 3.7).

Thus, the percent recovery of the NGS at 8 attomoles in the standard is $(7.3/8.0) \times 100 = 91.25\%$, which is the first row and last column of Table 3.7. The same procedure is followed for the 20-attomole reference in the standard equation to derive the 94% recovery of NGS. The results are in the second row, last column of Table 3.7. All of this is seen graphically in Figure 3.9, where one sees the NGS attomole values projected onto the horizontal axis, once one determines the RLU from the standard equation.

3.3   BRIEF INTRODUCTION TO BIOASSAY

Bioassay or biological assay is a scientific experiment conducted to measure the effects of a substance on a living organism and is a necessary procedure in the field of drug development and the monitoring of environmental pollutants or the monitoring of the effect on the environment of interventions such as pesticides or herbicides. Within biostatistics, bioassay is the class of procedures in which the amount of strength of an agent or stimulus is determined by the response of the subject. Usually, these are procedures by which the potency or the activity of a substance is estimated by studying its effects on living matter.

Bioassay is important for many product developments. Rahman, Choudhary, and Thomson (2001) give an excellent treatment of bioassay techniques for drug development. They discuss about 14 main requirements that a primary bioassay screen should meet. The text gives a fairly thorough overview of the various types of assay design from antiemetic assays to toxicity assays. The EPA has many useful monographs for bioassay assessment of contaminants in soil, air, and water samples.

In the field of pharmacology, the amount or measure of a drug activity required to produce an effect of given intensity is called potency. The more potent a drug, the lower the amount required to produce the effect. There are certain labels given to the term "potency." For example, assigned potency is the potency of a standard compound and estimated potency is the potency calculated from the assay. There are other labels as well. Relative potency refers to a comparison of the required quantities of two drugs – a standard (reference) drug and a new test drug to produce the same defined effect (response). One is, thus, validating the new or test versus the standard, much like we have been doing all along in method comparison. However, the statistical approach to this challenge is a bit more involved than our discussions thus far. Although we've been using the term assay, more formally there are two main types of assay one considers when determining potency and relative potency. Direct assays primarily involve a quantity known as individual effective dose (IED), which is defined as the amount of concentration or dose of the product to produce a particular response such as a defined reaction like death, cardiac arrest, or a certain level of a clinical parameter. This is a quantal or all or nothing response. In an indirect assay, one determines the magnitude of the response such as weight, time of survival, or level of toxicity, which may translate into a level of laboratory measurement such as the reticulocyte count, which we shall discuss later. In an environmental setting, one can determine the percent mortality of larvae over a range of doses of pesticide. In vaccine studies, various doses of the vaccine will result in various immunogenicity levels. This is a quantitative response. Here, specified doses are given to several subjects or animals and the nature of their response is recorded.

3.3.1   Direct Assay

Let's set up some notation. We are going to deviate from our linearity considerations and first introduce the direct assay we mentioned earlier. The purpose is to set the stage for notation and terminology relevant to bioassay in general. After discussion of the direct assay, we will return to linearity considerations and see how it applies to an indirect assay. In the direct assay, certain doses of the standard and test preparations are sufficient to produce a specified response, which can be directly measured. The ratio of the dose of the test preparation to the dose of the standard giving the specified response is an estimate of the potency, ρ. That is to say, if X_t and X_s are the doses of the test and the standard preparation or formulation producing the same effect or response, then the relative potency is defined as

$$\rho = \frac{X_s}{X_t}. \tag{3.15}$$

Thus, in such assays, the response or effect must be clear-cut and easily recognized.

A typical example of a direct assay is one for digitalis, in which a preparation of a compound such as Strophanthus is infused into an animal until the heart stops. The dose is immediately measured at that time. Table 3.8 shows a small experiment with eight animals on the test and eight animals on the standard preparation.

An estimate of the relative potency, ρ, is the ratio of the mean doses of the standard to the test or

$$R = \frac{\bar{x}_s}{\bar{x}_t} = \frac{1.88}{1.64} = 1.15. \tag{3.16}$$

Thus, 1 cc of the test preparation is approximately equal to 1.15 cc of the standard. The acceptance criteria for reproducibility of an assay are often defined by certain limit specifications. For example, according to the European Pharmacopoeia Commission, Council of European Directorate for the Quality of Medicines (EDQM, 1996), a test material passes if the estimated potency is not less than 80% and not more than 125% of the stated or standard potency and the 95% confidence limits of the estimated potency are not less than 64% and not more than 156% of the standard potency. Here, the mean test potency is less than the standard. Eighty percent of the standard potency is $0.80 \times 1.88 = 1.50$. Thus, the test potency of $1.64 > 1.50$ meets that criteria. The confidence interval (CI) criteria are confusing and not well stated. The 95% confidence interval on the mean test potency is $(1.4264, 1.8636)$. Sixty-four percent of the standard potency is $0.64 \times 1.88 = 1.20$. Clearly, the lower limit of the mean test CI is not lower than 1.20. It is worth noting here also that the CI on the difference on the means is $(-0.1061, 0.5736)$, which clearly contains 0, indicating no statistical difference between the two mean potencies.

Let's consider another approach. We reconsider the ratio of the potencies or the relative potency, ρ. An FDA (2001) criterion for bioequivalence (which we will discuss

TABLE 3.8 Data for Direct Assay

Strophanthus Preparation	Fatal Dose (0.01 cc/kg)	Mean
Standard	2.62, 1.87 1.67, 1.82, 1.61, 1.41, 1.99, 2.04	1.88
Test	1.57, 1.82 1.53, 1.67, 1.43, 1.23, 1.86, 2.05	1.64

in detail later) is that the 95% CI on the ratio of the means of two compounds must fall within the limits of 0.80–1.25. This is essentially saying that the test and the standard assays are the same or the two methods are comparable if the criteria are met. If we assume that criteria, or any criteria, to be our goal, then the challenge is to construct the 95% confidence limit for the relative potency, ρ, and determine if the criteria are met. In bioassay, this is done as a standard procedure. The most common approaches used are by Fieller (1954) and Bliss (1952). Both of these approaches are computationally intensive and require a computer program to develop the confidence interval. Both require the assumption of normality for both the test and standard samples. Fieller (1954) accommodates the possibility of a covariance structure or correlation between the standard and test sample. Bliss assumes that the samples are independent and a correlation structure is nonsignificant. Details of both these procedures are discussed in Hubert (1992). The correlation measures the strength of the linear relationship between two variables. Since our samples were taken independently, we assume nonsignificant correlation and use the approach by Bliss. The 95% CI for ρ is

$$(0.948, \quad 1.402). \tag{3.17}$$

So we can say that 1 cc of the test may have potency lying between 0.95 and 1.40 cc of the standard. Clearly, the lower limit is well above the limit of 0.80, but the upper limit of 1.25 is violated, and the test does not pass as being validated by this assay. The point being that one first sets the boundary criteria and then performs the assays, computes the relative potencies, and determines if the criteria are met. Clearly, from this example, depending on the criteria imposed by the regulatory agencies, the test assay may or may not be validated. Our purpose is not to set criteria, which may change given the circumstances under which the assay is conducted. Our goal is to examine the statistical procedures for the results needed to compare to the criteria, which is what we are doing.

3.3.2 Indirect Assay

Let's discuss the indirect assay and apply our linearity criteria. We are going to discuss a parallel line assay and demonstrate how to test for method validation of a test article versus the standard. We are going to look at an experiment much like that of Barth et al. (2008).

BALB/c mice of an appropriate age and strain are injected subcutaneously with preparations of a glycoprotein at three concentrations, once on Day 1 of the study. Study groups include three dilutions (50, 30, 10 IU) of a reference standard, and the same three dilutions of the test article for which potency is being assessed. The animals are observed until Day 5, at which time blood samples are collected and analyzed for reticulocyte counts. Method validation included the linearity of response and parallelism for the standard and the test article, and the potency of the test article relative to the standard. We shall describe all these procedures here. Remember, the purpose is to demonstrate validation of the test versus standard. This is essentially the method comparison strategy we have been discussing. There are 48 observations,

TABLE 3.9   Summary Data for Indirect Assay

Glycoprotein	Animals (N)	Dose IU/ml	Mean Log$_{10}$Dose	Mean Reticulocytes (10^5/mm^3)
Standard	8	10	1.0	10.54
Standard	8	20	1.30	7.21
Standard	8	40	1.60	3.97
Test	8	10	1.0	11.90
Test	8	20	1.30	8.59
Test	8	40	1.60	5.55

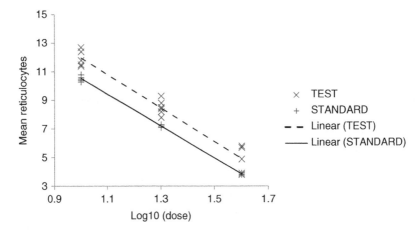

Figure 3.10   Indirect Assay Plot of Reticulocyte Count Versus Log10 (Dose)

eight animals in each of the three doses for each of the two preparations of the standard and the test. The data is summarized in Table 3.9. In our example, the doses are the same across the two preparations. Figure 3.10 is the plot of the reticulocytes against the Log10(Dose) for each of the preparations.

In order for the test to be considered for validation, three statistical criteria must be met. First of all, we must have a valid regression model. Secondly, we have parallelism of the dose–response lines, and thirdly, for our purposes here, they must satisfy linearity.

3.3.2.1   Examining the Three Criteria for Regression   If we assume that there is in fact a linear relationship between dose and response, let us set up some notation. Keep the glycoprotein/reticulocyte example in mind as we go through what follows. For the standard preparation we have the linear relationship

$$Y_{Si} = \beta_{0S} + \beta_S X s_i + e_i, \tag{3.18}$$

where Y_{Si} is the reticulocyte count for the ith individual mouse on the standard preparation and X_{Si} is the log dose to the base 10 for that individual. The term, e_i, is the error term associated with the measure of reticulocyte count, which is considered random and of course $i = 1, 2, \ldots, 24$. Likewise for the 24 mice on the test preparation, we have

$$Y_{Tj} = \beta_{0T} + \beta_T X_{Tj} + e_i \qquad j = 1, \ 2, \ \ldots, \ 24. \tag{3.19}$$

In a parallel model, we have the relationship of the log dose to response, as seen in Figure 3.10.

Computing the two lines separately, we have the predicted equation for the standard preparation as

$$\widehat{Y}_S = 21.193 - 10.900 \ X_S. \tag{3.20}$$

And for the test preparation, the predicted equation is

$$\widehat{Y}_T = 21.912 - 10.724 \ X_T, \tag{3.21}$$

where the hat (^) on both the Y's is a standard shorthand notation for the estimated or predicted value of the response from the data. It has the same interpretation as the value, Y_{Pred}, in equation (3.4).

It is unusual that the slopes are so closely aligned. This is not necessarily the case with many data situations. Thus, in Figure 3.10, the dose–response relationships of the two preparations are clearly distinguishable from each other. The rate of change of response (reticulocyte value) with the change in the log dose is about the same for both preparations, although the test preparation is consistently higher per log dose values compared to the standard. This is seen by the intercepts in equations (3.20) and (3.21) as well as in Figure 3.10. Again, the estimate of the two slopes, β_S and β_T, are approximately equal. This is an indication of parallelism, as is pictured in Figure 3.10. We have a formal statistical test for parallelism, which we will present later in this section. However, let's assume that parallelism does not exist and the lines appear as in Figure 3.11. Clearly, the rate of change of response to dose is different for both preparations. Stated differently, the preparation has two levels (Standard, Test) and the rate of change of response per dose for one level does not change at the same rate as the other level. The lines in Figure 3.11 are not parallel or the change in response is not consistent for the two preparations. This is called "interaction," while the plots in Figure 3.10 demonstrate no interaction.

We can formally test if interaction is present by examining what the statistical model looks like when interaction is present. It looks complicated but is easily explained. We'll go through the steps for testing for interaction and parallelism. Consider the following model that we state in general for our data. We take license with the notation to demonstrate the steps we go through.

$$Y = \beta_0 + \beta_1 \text{Log(dose)} + \beta_2 \text{Preparation} + \beta_3 [\text{Preparation} \times \text{Log(dose)}] + e. \tag{3.22}$$

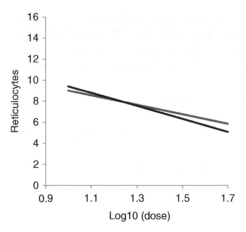

Figure 3.11   Example of Interaction

Here, Y is the response (reticulocytes). There are three possible factors influencing the response. They are the log(dose), preparation, and the possibility of an interaction expressed by the multiplicative term in equation (3.22) and as is seen in Figure 3.11.

We use an analysis of covariance (ANCOVA) model, much like the ANOVA model discussed in Chapter 2 to test for the significance of the interaction effect. Again, this effect is expressed by the multiplicative term, Preparation × Log(dose), one sees in equation (3.22). Covariance is a measure of how much two variables change together and how strong the relationship is between them. The variable, preparation, is a categorical variable and the log(dose) is a continuous variable often called a covariate. In an ANCOVA, we are in effect examining the influence of the preparation on the response while controlling for the effect of the log(dose) or the covariate. The data results are seen in Table 3.10.

The predicted or estimated model from the data in Table 3.10 is

$$Y = 21.912 - 10.724 \log(\text{dose}) - 0.719 \,\text{Preparation}$$
$$- 0.176 \,[\text{Preparation} \times \log(\text{dose})]. \tag{3.23}$$

So our first step is to determine if we have a significant interaction effect. We refer to the model in equation (3.22) and test for the interaction effect. This is simply testing the hypotheses:

TABLE 3.10   Summary Data for ANCOVA Model

Variable	Parameter	Estimate	p-Value
Constant	β_0	21.912	0.001
Log(dose)	β_1	−10.724	0.001
Preparation	β_2	−0.719	0.001
Interaction	β_3	−0.176	0.380

Test of Lack of Fit, p value $= 0.6277$.

$$H_0 : \beta_3 = 0 \quad \text{versus}$$
$$H_1 : \beta_3 \neq 0. \tag{3.24}$$

In Table 3.10, the p-value associated with this test is 0.380. We thus do not reject the null hypothesis, $H_0 : \beta_3 = 0$, and conclude that we have no significant interaction.

To see the parallelism, consider the following. Suppose we used the coding in our program to generate the model as preparation $= 0$ if the glycoprotein is the test and preparation $= 1$ if it is the standard. Then in (3.23), if preparation $= 1$, then the linear model for the standard preparation is

$$Y_S = 21.912 - 10.724 \ \log \ (\text{dose}) - 0.719 \times 1 - 0.176 \ (1 \times \log \ (\text{dose})) \tag{3.25}$$

or

$$Y_S = 21.193 - 10.900 \log(\text{dose}), \tag{3.26}$$

which is precisely equation (3.20).

Likewise, in equation (3.23), if preparation $= 0$ then the linear model for the test preparation is

$$Y_t = 21.912 - 10.724 \ \log \ (\text{dose}) - 0.719 \times 0 - 0.716 \ (0 \times \log \ (\text{dose}))$$

$$\text{or} \quad Y_t = 21.912 - 10.724 \ \log(\text{dose}), \tag{3.27}$$

which is precisely equation (3.21).

Note that equations (3.26) and (3.27) have approximately the same slope, but different intercepts. Thus, they are parallel, as seen in Figure 3.10. Therefore, one sees if you do not reject the null hypotheses that the coefficient or beta for the interaction term is equal to zero, then you in effect demonstrate parallelism as the interaction term disappears. The only thing left to demonstrate is that the linear model is appropriate. As we stated earlier, one can examine the residuals as in Figure 3.12 to see that the plot looks reasonable for the linear fit of the parallel model (3.25). Also note in Table 3.10 that a "Lack-of-Fit" test was performed with a p-value $= 0.628$. The lack-of-fit test is a test of how well the model fits the data. It works particularly for data that is replicated, as is often the case for dose levels in the type of assay we are discussing. A p-value of 0.05 or less is a significant lack of fit, and one would reject the model as being appropriate. The null hypothesis is that the model fits the data well versus the alternative that it does not. Our p-value $= 0.628$, which indicates we would not reject the null hypothesis, and this clearly demonstrates that the model is appropriate. Most statistical packages have this option. Also, recall that we want to examine the residual plot for the appropriateness of the linear fit. Figure 3.12 shows a good random scatter of the residuals about their mean $= 0$ line indicating the appropriateness of the linear fit. Thus, to summarize, we have the lack-of-fit test for confirming

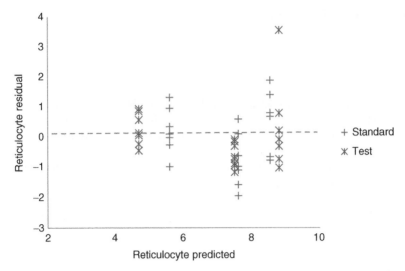

Figure 3.12   Indirect Assay Parallel Model Plot of Residuals Versus Predicted Values

the appropriateness of the regression model, the rejection of the interaction term in the full model to indicate parallelism, and a reasonable residual plot in the parallel model to indicate linearity. Our next goal is to determine relative potency and its 95% confidence limits.

3.3.2.2   *Point Estimate of the Relative Potency, ρ*   Having established that the statistical requirements are now met for conducting the analysis of the parallel line assay, we proceed to discuss the relative potency. For our indirect assay, let $\hat{\rho}$ be the estimate of the relative potency and let

$$Q = \log \ \hat{\rho}, \tag{3.28}$$

then Q is defined as

$$Q = \frac{\overline{Y}_t - \overline{Y}_s}{b} - (\overline{X}_t - \overline{X}_S). \tag{3.29}$$

Here, $\overline{Y}_t$ is the mean of the test response from the last column of Table 3.9 or 8.68 and $\overline{Y}_s$ is the mean of the standard response or 7.24. $\overline{X}_t$ and $\overline{X}_S$ are the means of the log doses of the test and standard preparations, respectively, from the fourth column of Table 3.7. Since this is a symmetric assay, that is, one in which dose levels vary in the same manner for both the test and the standard, then $\overline{X}_t = \overline{X}_S$ (Table 3.9) and thus $\overline{X}_t - \overline{X}_S = 0$. The value, b, is the estimate of the common slope of the parallel lines.

If b_s is the estimate of the slope, β_S, of the line for the standard preparation (3.18) and b_t is the estimate of the slope, β_t, of the line for the test preparation (3.19) and assuming parallelism or $b_s = b_t = b$, then the best choice of b is

$$b = \frac{w_t b_t + w_s b_s}{w_t + w_s}. \tag{3.30}$$

This is a weighted average of the two slope estimates. The weights, $w_k(k = s, t)$, are usually the inverse variance (var) of the estimate, b_k, that is,

$$w_k = \frac{1}{\text{var}(b_k)} = \frac{\sum (X_k - \overline{X}_k)^2}{\text{MSE}}, \quad k = s, t. \tag{3.31}$$

Here the MSE can be taken as the mean-squared error result in the ANOVA for the model, equation (3.22) (Hubert 1992). At any rate, one will notice that the MSE will cancel in both the numerator and denominator of equation (3.30). For symmetrical assays as we have here with dose levels varying in the same manner for both the test and the standard, then $w_t = w_s$ and thus,

$$b = \frac{b_t + b_s}{2}. \tag{3.32}$$

Thus, from equations (3.26) and (3.27) in our aforementioned example, we have

$$b = \frac{(10.724 + 10.900)}{2} = 10.812. \tag{3.33}$$

The point estimate $Q = \log \hat{\rho}$ from equation (3.29) is

$$Q = \frac{8.679 - 7.241}{10.812} = 0.1331. \tag{3.34}$$

The point estimate of ρ is the antilog of Q or $\text{anti}\log(Q) = 10^{0.1331} = 1.3586.$
$$\tag{3.35}$$

Thus, 1 unit of the test preparation is approximately equal to 1.36 units of the standard preparation. So the estimated potency of the test is certainly more than 125% of the standard. If these criteria were in place, then the assay would not be validated and we do not achieve method validation. However, we take it a step further and compute the 95% confidence interval on the relative potency. We'll assume that the two preparations are independent. The confidence interval formula for $Q = \log \hat{\rho}$ is rather cumbersome, and although it is usually computer calculated, we present it here for the sake of completeness, and also because it uses many of the elements we've discussed already and one can actually hand compute it.

The confidence limits for the log relative potency of Q are from Hubert (1992), that is,

$$\frac{Q \pm t \frac{s}{b} \sqrt{\frac{Q^2}{D} + (1 - g)\left(\frac{1}{n_s} + \frac{1}{n_T}\right)}}{1 - g}. \tag{3.36}$$

Q is of course is 0.1331. The variable, t, is the Student's t-critical value on $n_s + n_T - 2$ degrees of freedom (df), where, of course, n_s and n_T are the sample sizes of the standard and test preparations, which in our case is 24 for each. Thus, the critical t-value for $24 + 24 - 2 = 46$ df is 2.012. S^2 is the MSE (mean-squared error or residual variance) from the ANOVA (Chapter 2) for the model equation (3.22), which is 0.115. Thus, $S = \sqrt{\text{MSE}} = 0.339$. The parameter, b, is from equation (3.33) and is equal to 10.812. D is defined as the sum of the corrected sums of squares of the log(dose), which in our case is

$$D = \sum (X_s - \overline{X}_s)^2 + \sum (X_T - \overline{X}_T)^2. \tag{3.37}$$

We change notation slightly here to use cap T for the test so as not to confuse it with the critical Student's t-value. Recall we use the same dose schedule for both the standard and the test preparation. So from Table 3.9 (Mean Log10 Dose column) and the data for this assay, we have

$$D = \sum (X_s - \overline{X}_s)^2 + \sum (X_T - \overline{X}_T)^2 = 0.18 + 0.18 = 0.36. \tag{3.38}$$

The value g is

$$g = \frac{t^2 s^2}{D\, b^2}. \tag{3.39}$$

All of these values are defined earlier and thus doing the calculations, $g = 0.01105$ and $1 - g = 0.9889$. Thus, the 95% confidence interval for the log relative potency, Q, is

$$(0.112, \quad 0.158). \tag{3.40}$$

Taking the antilog of both these limits ($10^{0.112}$, $10^{0.158}$), the 95% confidence interval for the relative potency is

$$(1.294, \quad 1.439). \tag{3.41}$$

So we can say with 95% confidence that 1 unit of the test preparation may have potency lying between 1.29 and 1.44 units of the standard. Recall the relative potency is 1.36.

We have been discussing parallel line assays in which the dose response has been linear. There are more complicated models, obviously, in which the relationship is nonlinear. One such relationship is noted in Figure 3.13. Here the relationship is sigmoidal, and the underlying function characterizing the shape of the curves is the logistic function (Gottschalk and Dunn 2005). The mathematics are quite involved using the multiparameter logistic model. However, there is a test of parallelism as well as computation of relative potency with confidence limits. A software package is certainly needed in this case. The data in Figure 3.13 is motivated by the desire to examine the relative potency of two agents (test vs. reference). There were 17 dose concentrations with three observations per dose for each agent, for a total of 102 total data points. The test of parallelism yielded a nonsignificant p-value or a

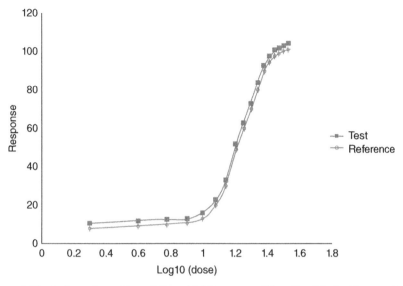

Figure 3.13 Indirect Assay Plot of Sigmoidal Response of Test (line filled with square) and Reference (line filled with diamond) Versus Log10 (Dose)

nonsignificant interaction term, indicating strong evidence for parallelism, as seen in Figure 3.13. The relative potency of the test to the reference was 1.01. Thus, 1 unit of the test preparation is approximately equal to 1.01 units of the standard preparation. For a thorough discussion of nonlinear parallel line assays, the reader is referred to Gottschalk and Dunn (2005). If one were to use the criteria that the estimated potency is not less than 80% and not more than 125% of the stated, then a relative potency of 1.01 would certainly validate the test agent.

3.4 SENSITIVITY, SPECIFICITY (SELECTIVITY)

In a clinical setting, one is interested in assuring that a diagnosis is made correctly. That is to say, clinically, one wishes to maximize the probability that a positive test for a disease is in fact positive, and likewise that a negative test is in fact negative. Likewise, in our laboratory context, we are comparing one device or equipment (new) to another (standard) to validate the performance of the new device. We assume that the endpoint is to be diagnostic (positive or negative). How well does the new equipment perform compared to the standard? We are again in the context of method validation. However, now our response is categorical. In the previous sections, we considered data that had a continuous response such as laboratory values. Now we merely want to know if something is present or not, and we want to determine it with high accuracy.

Table 3.11 gives a general set up. One can see that according to the Standard or Reference procedure (usually a confirmed pathological diagnosis) that the number of true positives is $a + c$ and the number of true negatives is $b + d$. We always want to know how accurate the diagnostic test is in predicting disease or no disease. Specificity is

TABLE 3.11   Setup for Sensitivity Specificity Analysis

New Device	Standard Device		Total
	Disease Present	Disease Absent	
Negative	a	b	$a + b$
Positive	c	d	$c + d$
Total	$a + c$	$b + d$	$a + b + c + d$

Sensitivity $= c/(a + c)$; specificity $= b/(b + d)$.

the chance that the individual has no disease when the diagnostic (new) test registered negative. That is true negative. That value is $b/(b + d)$. Sensitivity is the chance that the individual has the disease when the diagnostic test registered positive, that is, true positive. That value is $c/(a + c)$. An actual data example will make this clearer.

We refer to Burggraf et al. (2005), in which they attempted to test a sensitive and specific assay for reliable and flexible detection of members of the Mycobacterium tuberculosis complex (MTBC) in clinical samples. They tested a new nucleic acid technique (NAT) against a standard to determine its reproducibility of results. They thus compared the two NATs, COBAS AMPLICOR (standard) to the newer large Volume LightCycler Assay, which was considered more cost-effective. While culture is still the gold standard for specific diagnosis of TB, detecting members of MTBC may require several weeks to produce results. NATs usually provide results within a day. The procedure is well presented in their 2005 article.

At the time of the writing of Burggraf et al. (2005), the only NAT-based assays approved by the FDA were the Amplified *Mycobacterium tuberculosis* Direct Test (MTD; Gen-Probe, San Diego, California.) and the COBAS AMPLICOR *M. tuberculosis* assay (Roche Diagnostics, Mannheim, Germany). Given well-established clinical samples, the COBAS performed well with smear-positive respiratory samples (Reischl et al. 1998). However, the flexibility of this assay was limited, since it relied on a relatively slow block cycler amplification process and time-consuming colorimetric detection of amplification products. For economic reasons, most laboratories can make only one PCR run per day; thus, just-in-time testing of single samples is not feasible.

Thus, in their comparison, Burggraf et al. (2005) reached the overall conclusions that the LightCycler assay is certainly comparable to the COBAS with sensitivity around 100% and specificity around 97% in the TB samples examined. It was considered more cost-effective and faster. The article does describe some operational limitations not discussed here. As a demonstration of this rather simple technique, consider Table 3.12 in comparing a new device to standard in detecting a disease. Clearly, the sensitivity is 95/102 or 93% and the specificity is 93/98 or 95%. Other statistics of interest might be the positive predictive value, which is the number of true positives out of those predicted to be positive by the new device. This is merely $c/(c + d)$ in Table 3.11 and $95/100 = 95\%$ in Table 3.12. Likewise, one might be interested in the negative predictive value, which is the new device's ability to correctly predict a negative result. This would be from the first row of Table 3.11 or $b/(a + b)$, which is $93/100 = 93\%$ in Table 3.12.

TABLE 3.12 Numerical Example of Sensitivity and Specificity

New Device	Standard Device		Total
	Disease Present	Disease Absent	
Negative	7	93	100
Positive	95	5	100
Total	102	98	200

Sensitivity $= 95/102 = 0.93$; specificity $= 93/98 = 0.95$.

Just as we learned in Chapter 2, we can place confidence intervals on the population value of sensitivity and specificity. This is a simple formula for a single proportion and can be done easily by hand. We construct it here. The sensitivity is $95/102 = 0.93$. Let the estimate $\hat{p} = 0.93$. Thus, $\hat{q} = 1 - \hat{p} = 0.07$. The standard error (SE) of this estimate is $\sqrt{\hat{p}\hat{q}/n}$. In our case, $n = 102$ and SE $= 0.0253$. The 95% confidence interval for the population value of the sensitivity based on this data is, thus, defined:

$$\hat{p} \pm z_{\alpha/2} \text{ (SE)}, \tag{3.42}$$

where $z_{\alpha/2} = 1.96$, which is the critical value of a normal distribution needed for a two-sided 95% confidence limit. Putting this all together, we have the 95% confidence limit of sensitivity as

$$(0.880, \ 0.980). \tag{3.43}$$

Sometimes, one likes to geometrically get an idea of the accuracy of a test versus the standard. This is done by examining what is known as a receiver operating characteristic (ROC) curve. It shows the trade-off between the sensitivity and specificity (any increase in sensitivity will be accompanied by a decrease in specificity). The plot is actually the true positive rate (sensitivity) versus the false positive rate (1-specificity). One then measures the area under the curve (AUC) to determine the accuracy of the test result; this is best seen by example. Table 3.12 is only one table with one specificity and one sensitivity. However, one can look at the observed data plus the extreme values and construct the curve using three points as follows: If the sensitivity is 0 and the specificity is 1 (i.e., $c = 0$ and $b = 98$ in Table 3.11 and Table 3.12), then 1-specificity $= 0$. The observed sensitivity is 0.93 and observed specificity is 0.95. Thus, 1-specificity $= 0.05$. The last possibility is the sensitivity $= 1$ and specificity $= 0$ (i.e., $c = 102$ and $b = 0$ in Tables 3.11 and 3.12). Thus, 1-specificity $= 1.00$. These points are plotted in Figure 3.14. The AUC $= 0.940$ or the test or new device has a 94% accuracy. To fully understand the interpretation of the AUC, the total area of the grid in Figure 3.14 is equal to one. The AUC $= 0.94$. An AUC $= 1$ would be a perfect test. Thus, one can see that the more AUC, the greater the accuracy of the test. An AUC of 0.50 or less is often considered a test of little or no worth. The computation of the AUC is rather complicated mathematically, and is usually done by computer software.

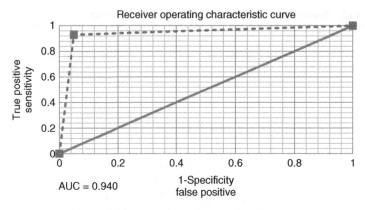

Figure 3.14 ROC Curve for Data in Table 3.12

Now, let's examine a dose–response analog of a dilution experiment much like that of Burggraf et al. (2005). We examine a dilution series of DNA from *M. tuberculosis* series H37, which was assayed on an instrument in 20 μL capillaries. Five experiments were performed in parallel for each dilution. The result is shown in Table 3.13. At each DNA concentration/dilution, we have the number of positive results out of the five experiments. The ROC is given in Figure 3.15. The ROC is constructed from this series of experiments. To demonstrate some of the data, note that in Table 3.13 there are 22 positive results and 18 negative results. These totals are sort of the gold standard for constructing the ROC and determining sensitivity and specificity sequentially from the experiments. For example, at DNA = 50 genomes/μL, assuming an accurate instrument, we have a cumulative (from the top) total of nine positives, thus, inferring a sensitivity at that point of 9/22 = 0.4091. We have one negative result inferring that the true negatives from that point must be 17 or a specificity of 17/18 = 0.9444. This is 1-specificity = 1 − 0.9444 = 0.0556. Thus, in Figure 3.14, we see that at a false positive rate of 0.0556, the true positive rate or sensitivity is 0.4091. At the end of Table 3.13 (last row), we have accumulated 22 positives for a sensitivity of 22/22 = 1.0. The specificity at that point is the remaining negatives or 5/18 = 0.2778. Thus, 1-specificity = 1−0.2778 = 0.7227. The

TABLE 3.13   Sensitivity of M TB DNA Detection

DNA Concentration Genomes/μl	Number of Positives from 5 runs
100	5/5
50	4/5
30	4/5
10	3/5
5	3/5
2.5	2/5
1.0	1/5
0.1	0/5

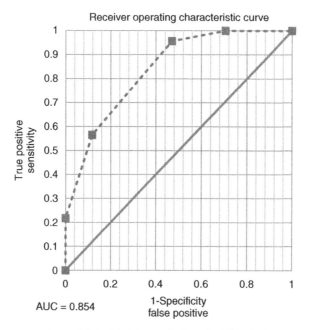

Figure 3.15 ROC Curve for Data in Table 3.13

final plotted point on the ROC is (0.723, 1.00). Note the AUC $= 0.854$ or the test has about an 85% accuracy.

Thus, an ROC curve demonstrates several things: It shows the trade-off between sensitivity and specificity (any increase in sensitivity will be accompanied by a decrease in specificity). The closer the curve follows the left-hand border and then the top border of the ROC space, thus, occupying more space (AUC) on the square grid as in our plots, then the more accurate the test. The closer the curve comes to the 45° diagonal or AUC $= 0.50$ of the ROC space, the less accurate the test. Regarding the AUC, this is a measure of accuracy and is called the area under the ROC curve. An area of 1 represents a perfect test; an area of 0.5 represents a very poor test in terms of accuracy. Some individuals use a rough guide for classifying the accuracy of a diagnostic test like a traditional academic point system: that is, $1.00 =$ perfect, $0.90–0.99 =$ excellent, $0.80–0.89 =$ good, $0.70–0.79 =$ fair, $0.60–0.69 =$ poor, and 0.59 or less $=$ fail.

3.5 METHOD VALIDATION AND METHOD AGREEMENT – BLAND-ALTMAN

As one can see by now, method validation in a sense is truly a technique of deter-mining agreement between two methodologies. We've taken a linearity approach in part to address the issue, but it is not enough. Let's revisit Table 3.1 and Figure 3.2.

Recall the $R^2 = 0.9965$ or the variation in the Test method explained by the Standard is about 0.9965 or, expressed as a percent, is about 99.65%. Recall also that we had a significant linear fit, $p = 0.001$, or a significant association between the two methods, and the line went through the origin. While all these statistics indicated a strong linear relation between the two methods, this does not translate into a strong case for agreement of the two methods, unless each method gave the exact same value for each unit measured. The R^2 and thus the correlation measure the strength of a relation between two variables, not the agreement between them. We have perfect agreement only if the points in Figure 3.2 follow along the line of equality, but we will have perfect correlation if the points lie along any straight line. For example, if each Y variable is exactly twice the value of each X variable, one can generate a perfectly straight line where all the (x, y) points lie along the line and the correlation would be perfect and the $R^2 = 1$, but obviously, the two methods would not be in strong agreement, as one gives twice the result of the other. Thus, the R^2 and correlation can indicate strong association or near-perfect prediction of one variable based on the value of the other, but as we've seen here, it does not translate into agreement of the two methods. It is unlikely that different methods will agree exactly by giving identical results for all individuals. We thus ask by how much one method is likely to differ from the other? If it is not enough to cause problems in clinical or laboratory interpretation, then we can replace one method by the other. So while all the methodologies we discussed earlier give good statistical and graphical representations of accuracy and degrees of association, what is yet another approach that we might consider to measure true agreement and the boundaries or criteria that one considers for true agreement?

Perhaps the most popular approach to this method of agreement has been taken by Bland and Altman (1986). Their methodology, which is basically graphical, plots the difference between the two methods for each observation on the vertical axis versus the average of the two values on the horizontal axis. We do this with our data in Table 3.1, where for each subject we plot (TEST−STANDARD) on the vertical axis and (TEST + STANDARD)/2 on the horizontal axis (Figure 3.16). To come to some conclusion as to the viability of agreement between the two methods, we compute some statistics.

Note the mean difference, $d = 0.0056$. This is the solid line in the center of Figure 3.16. The upper and lower dashed lines are the 95% confidence interval on the mean $(−0.041, 0.042)$. The standard deviation is $S = 0.0843$. We examine the limits, that is, 2 standard deviations below the mean and 2 standard deviations above the mean:

$$d - 2S = 0.0056 - 2(0.0843) = -0.1629$$

$$d + 2S = 0.0056 + 2(0.0843) = 0.1741. \tag{3.44}$$

Thus, $(d - 2S, d + 2S) = (-0.163, 0.174)$. Note that the quantity, $d \pm 2S$, is sometimes referred to as the bias. The mindset is often that if the computed differences fall within the interval, $(d - 2S, d + 2S)$, then we can use the two methods interchangeably. Note that in Figure 3.16, all the differences do in fact fall within the limits. Notably, all the difference values except one fall within the interval $(-0.10, 0.10)$.

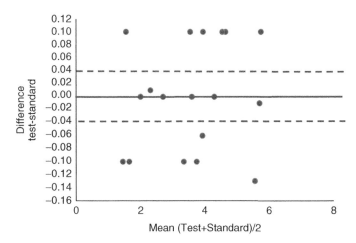

Figure 3.16   Bland–Altman Plot for Data in Table 3.1

According to Vannest et al. (2002), if the values are within 10% of each other, then the bias between the methods is minimized. This choice of 10% is usually predicted upon the indication or analyte being examined. The value, 10%, is just one choice that may be acceptable. Upon a closer look at Figure 3.16, we see that most of the values, actually 10, are at the −0.1 or 0.10 mark. However, note that one point to the far lower right of Figure 3.16 has the value, −0.13, which is outside the −0.10 limit. It's only one point and should not be bothersome. However, the Bland–Altman approach allows us to take yet another closer look at the limits, $(d - 2S, d + 2S)$. They put a 95% confidence interval on the limits, $d - 2S$ and $d + 2S$, which we construct as follows.

The standard error (SE) of $d - 2S$ as well as $d + 2S$ is $(3 * S^2/N)^{1/2}$. Recall in our sample, $N = 18$ and $S = 0.0843$. Thus, SE $= (3 * 0.0843^2/18)^{1/2} = 0.0344$. The t-value for a 95% confidence interval on $N - 1 = 17$ degrees of freedom (df) $= 2.1098$. So the 95% confidence interval on $d - 2S$ is

$$d - 2S - 2.1098(0.0344) \text{ and } d - 2S + 2.1098(0.0344)$$

or

$$(-0.1629 - 0.0726, \ -0.1629 + 0.0726) = (-0.236, \ -0.090). \tag{3.45}$$

Likewise, the 95% confidence interval for $d + 2S$ is

$$d + 2S - 2.1098(0.0344) \text{ and } d + 2S + 2.1098(0.0344) \tag{3.46}$$

or

$$(0.1741 - 0.0726, \ 0.1741 + 0.0726) = (0.102, \ 0.247).$$

Note that the difference, −0.13, in Figure 3.16 is well within the interval (−0.236, −0.090), that is, it is not below the lower limit of −0.236 and is thus not significantly outside the bias range as defined here. None of the differences in the top half of Figure 3.16 are above the limit of 0.247; thus, one can feel fairly comfortable of acceptable agreement between the two methods. In the next chapter, we shall discuss how outliers can be detected using the Bland–Altman procedure we discussed here.

REFERENCES

Barth T, Oliveira PR, D'Avila FB and Dalmora, SL. (2008). Validation of the normocythemic mice bioassay for the potency evaluation of recombinant human erythropoietin in pharmaceutical formulations. Journal of AOAC International 91(2): 285–291.

Bland JM and Altman DG. (1986). Statistical methods for assessing agreement between two methods of clinical measurement. Lancet 327(8476): 307–310.

Bliss CI. (1952). The Statistical Method in Bioassay, Hafner Publishing Company, New York.

Burggraf S, Reischl U, Malik N, Bollwein M, Naumann L and Olgemöller B. (2005). Comparison of an internally controlled, large-volume LightCycler assay for detection of *Mycobacterium tuberculosis* in clinical samples with the COBAS AMPLICOR assay. Journal of Clinical Microbiology 43(4): 1564–1569.

Curiale MS, Sons T, McIver D, McAllister S, Halsey, VB, Roblee D and Fox TL. (1991). Dry rehydratable film for enumeration of total coliforms and *Escherichia coli* in foods: collaborative study. Journal of Association of Official Analytical Chemists 74(4): 635–648. Available at http://www.3m.com/intl/kr/microbiology/regu_info/9.pdf.

European Pharmacopoeia Commission, Council of Europe, European Directorate for the Quality of Medicines European Pharmacopoeia (EDQM, 1996) 3rd ed.

Fieller, EC. (1954). Some problems in interval estimation. Journal of the Royal Statistical Society, B 16 (2): 175–185. JSTOR 2984043.

Gottschalk PG and Dunn JR. (2005). Measuring parallelism, linearity and relative potency in bioassay and immunoassay data. Journal of Biopharmaceutical Statistics 15: 437–463.

Hall TG, Smukste I, Bresciano KR, Wang Y, McKearn D and Savage RE. (2012). Identifying and overcoming matrix effects in drug discovery and development, tandem mass spectrometry – applications and principles; in J Prasain (Ed.), InTech, ISBN: 978-953-51-0141-3, Available from: http://www.intechopen.com/books/tandem-mass-spectrometry-applications-andprinciples/identifying-and-overcoming-matrix-effects-in-drug-discovery-and-development.

Hubert JJ. (1992). Bioassay. 3rd ed., Kendall/Hunt Pub, Dubuque, Iowa.

Mandel J and Stiehler RD. (1954). Sensitivity – a criterion for the comparison of methods of test. Journal of Research of the National Bureau of Standards 53(3): 155–159.

Rahman A, Choudhary MI and Thomson WJ. (2001). Bioassay Techniques for Drug Development. Harwood Academic Publishers, Singapore.

Reischl U, Lehn N, Wolf H, and Naumann L. (1998). Clinical evaluation of the automated COBAS AMPLICOR MTB assay for testing respiratory and nonrespiratory specimens. Journal of Clinical Microbiology 36: 2853–2860.

U.S. Department of Health and Human Services, Food and Drug Administration. Guidance for Industry on Bioanalytical Method Validation, May 2001, Available from: www.fda.gov/downloads/drugs/guidances/ucm070107.pdf.

Vannest R, Douglas J. Markle V, Bozimowski D. (2002). Abbott Laboratories: Irving TX. Evaluation of Chemistry Assays on the Abbott ARCHITECT c8000™ Clinical Chemistry System Presented at the 54th Annual Meeting of the American Association of Clinical Chemistry. July–August.

4

METHODOLOGIES IN OUTLIER ANALYSIS

4.1 INTRODUCTION

Outliers are considered extreme observations and may in fact be legitimate data points unless there are documented problems with instrumentation, observers, or exterior conditions. Statisticians, as a rule, discourage omitting data points from an analysis because one believes they are outliers. Outliers have been defined in various ways. An outlier may be defined as an observation that "appears" to be inconsistent with other observations in the data set or experiment. An outlier can be defined statistically as having a low probability that it belongs to the same statistical distribution as the other observations in the data set. On the other hand, we will see by example that an extreme value is an observation that might have a low probability of occurrence but cannot be shown statistically or graphically to be derived from a different distribution than the remainder of the data.

Statistical outlier detection has become a timely topic in all experimental approaches as a result of the US Food and Drug Administration's out of specification (OOS) guidance and increasing emphasis on the OOS procedures of pharmaceutical companies. When a test fails to meet its specifications, the initial response is to conduct a laboratory investigation to seek a probable cause. As part of that investigation, an analyst looks for an observation in the data that could be classified as an outlier. Ordinarily, a chemical result cannot be omitted with an outlier test according to the FDA guidance "Investigating Out of Specification (OOS) Test Results for Pharmaceutical Production" (FDA 2006) and the US Pharmacopeia (2006).

Introduction to Statistical Analysis of Laboratory Data, First Edition.
Alfred A. Bartolucci, Karan P. Singh, and Sejong Bae.
© 2016 John Wiley & Sons, Inc. Published 2016 by John Wiley & Sons, Inc.

There are several obvious reasons for wanting to detect and explain outliers. They can provide useful information about the process. The process might have shifted an outlier, and this can arise due to a shift in the location (mean) or in the scale (variability) of the process. Though an observation in a particular sample might be an outlier candidate, one should be aware again that the process might have shifted. Sometimes, the unusual result is a recording error or a measurement error.

Outliers also come from incorrect specifications that are based on the wrong distributional assumptions at the time the specifications are generated. However, the distributional assumptions used should be known ahead of time, as should be the measurement systems for the process being measured.

Although the primary focus of this chapter is the detection of outliers, what one does with the outlier can lend itself to several possibilities. One often takes the approach that once an observation is identified, by means of graphical or visual inspection, as a potential outlier, some cause analysis should begin to determine whether a probable cause can be found for the unusual or spurious result. This approach is not entirely acceptable. One needs to formally, through acceptable statistical evaluation, determine if this is a mathematically justified outlier by definitions, which we will make clear. If no justifiable cause can be determined, and a retest can be justified, the potential outlier should be recorded for future evaluation as more data become available. When reporting results, it is prudent to report conclusions with and without the suspected outlier in the analysis. This provides some sensitivity as to the influence of the suspected outlier on the outcome. Removing data points without a reasonable cause is insufficient. Robust or nonparametric statistical methods are the methods for analysis that we shall discuss. When we refer to a robust method, we mean a statistical technique that performs well even if its assumptions are somewhat violated by the true model from which the data were generated. We will use this term when appropriate in our discussions later.

Thus, there are various approaches to outlier detection, depending on the application and number of observations in the data set. Iglewicz and Hoaglin (1993) provide a comprehensive approach to labeling, accommodation, and identification of outliers. Visual inspection alone cannot always reliably identify an outlier and can lead to misclassifying an observation as an outlier. Because parameters used in estimation, such as the mean, may be highly sensitive to outliers, statistical methods were developed to accommodate outliers and to reduce their impact on the analysis. Some of the more commonly used detection methods are discussed in this chapter.

4.2 SOME OUTLIER DETERMINATION TECHNIQUES

We consider a few techniques for detection of outliers, keeping in mind they can only alert us to problems that may or may not exist. Some outlier or extreme observation techniques include the following:

- Grubb's test (rather crude)
- Use of Studentized range

- Sequential test of many outliers
- Mahalanobis distances
- Dixon Q-test
- Box plot
- Nonparametric approaches
- Combined method comparison and outlier detection using linearity
- Combined method/outlier comparison using the Bland–Altman procedure
- Outlier variance techniques.

We demonstrate each by an example.

Table 4.1 consists of total laboratory cholesterol (TC) readings from 20 subjects with hypercholesterolemia. Three instruments were used to record the TC values for each subject. We will use the data from this table for our next several demonstrations of the outlier detection techniques.

TABLE 4.1 Sample Data

	Total Cholesterol Readings		
Subject	Method A	Method B	Method C
1*	266	270	265
2	273	277	272
3	273	277	272
4	274	277	274
5	275	279	274
6	276	280	277
7	276	281	277
8*	276	285	277
9	277	280	277
10	277	279	276
11	278	280	278
12	278	281	278
13	278	282	279
14*	279	289	278
15	279	283	280
16	280	284	279
17	281	285	279
18	282	285	281
19	282	286	280
20	284	288	285
Minimum	266	270	265
Maximum	284	289	285
Mean	277.2	281.4	276.9
Standard deviation	3.995	4.453	4.141
Standard error	0.893	0.996	0.926
N	20	20	20

4.2.1   Grubb Statistic

The Grubb test, also known as the maximum normalized residual test, can be used to test for outliers in a single data set. We are careful to note that this test assumes normality, so one must test the data for normality before applying the Grubb test. If one is not certain that the data is normal, then there are formal statistical procedures for doing so. A statistician should be consulted in this case.

The Grubb test detects one outlier at a time. For multiple outliers, there are alternative procedures. However, if one wishes to use the Grubb test and suspects multiple outliers, then one can iterate the procedure, that is, delete the single outlier detected on the first run and run the Grubb test again. Repeat this process until no outliers are detected. This is an easy procedure and can be done fairly quickly by hand with the appropriate table of critical values. We consider the data for Method A in Table 4.1 as we discuss this procedure.

More formally, the Grubb test can be defined by the following hypotheses:

H_0: There are no outliers in the data set.

H_1: There is at least one outlier in the data set.

The test statistic is

$$G = \frac{|(X_{i(\text{extreme})} - \overline{X})|}{S}, \tag{4.1}$$

where $\overline{X}$ is the mean and S is the standard deviation. $X_{i(\text{extreme})}$ is the most extreme value from the mean, either the minimum or maximum value of a sample of size N, $i = 1, \ldots, N$. The symbol, $||$, of course, denotes the absolute value. The statistic, G, takes the most extreme value from the mean, either the minimum or maximum value in the data, and subtracts the mean from it. Now let's refer to Table 4.1, Method A data. Consider the max and min values and determine which is most extreme from the mean.

In Method A, the minimum = 266 is further from the mean, 277.2, than the maximum = 284. The Grubb Statistic is

$$G = \frac{|(266 - 277.2)|}{S} = \frac{|(266 - 277.2)|}{3.995} = \frac{11.2}{3.995} = 2.803.$$

The value, 2.803, is compared to the critical value of the Grubb statistic for a sample of size = 20, which for the 5% significance level is equal to 2.56 for a one-sided test or 2.71 for a two-sided test. The value 2.803 > 2.56 and 2.803 > 2.71, so the value 266 is considered an outlier for either the one-sided or two-sided test. Look at Table 4.2 for a sample of Grubb critical values, and one can see that these numbers apply at row, sample size $N = 20$. Although the table starts at a sample size of $N = 3$, a strong note of caution is that this test should not be used for sample sizes of six or less since with a sample size that small, it frequently tags most of the points as outliers.

TABLE 4.2   Example of Grubb Table of Critical Values

N	Grubb Critical Value Table			
Alpha Level	0.10	0.05	0.025	0.01
3	1.15	1.15	1.15	1.15
4	1.42	1.46	1.48	1.49
5	1.60	1.67	1.71	1.75
6	1.73	1.82	1.89	1.95
10	2.03	2.18	2.29	2.41
15	2.25	2.41	2.55	2.71
20	2.38	2.56	2.71	2.88
25	2.49	2.66	2.82	3.01
30	2.56	2.75	2.91	3.10
35	2.63	2.82	2.98	3.17
40	2.68	2.87	3.04	3.24
45	2.73	2.92	3.09	3.29
50	2.77	2.97	3.13	3.34
55	2.81	2.99	3.16	3.38
60	2.84	3.03	3.20	3.41
65	2.87	3.06	3.23	3.44
70	2.90	3.09	3.26	3.47
75	2.92	3.12	3.28	3.50
80	2.95	3.14	3.31	3.52
85	2.97	3.16	3.33	3.54
90	2.99	3.18	3.35	3.56
95	3.01	3.19	3.36	3.58
100	3.02	3.21	3.38	3.60
110	3.06	3.24	3.42	3.63
115	3.07	3.26	3.43	3.65
120	3.09	3.27	3.44	3.66

One notes in Table 4.2 that the sample sizes after six are not in units but multiples of five. One can either interpolate to obtain a critical value or use the formula

$$\frac{N-1}{\sqrt{N}} \sqrt{\frac{t^2_{\frac{\alpha}{N},N-2}}{N-2+t^2_{\frac{\alpha}{N},N-2}}}, \tag{4.2}$$

where t is the t-statistic, α is the alpha level, and N is the sample size. The degrees of freedom are $N-2$. For example, solving equation (4.2) the critical value for $N=37$ at $\alpha=0.05$ (or $\alpha/N=0.05/37=0.00135$) is 2.84. One can see here that deriving a critical value of a t-statistic for an alpha level $=0.00135$ is not easily done from a standard t-table. The t-values for a specified alpha level and degrees of freedom can be easily computed from a standard computer program.

Alternatively, interpolating from Table 4.2 the critical value, Y, for $N = 37$ in the 0.05 column is computed as

$$\frac{37 - 35}{40 - 35} = \frac{Y - 2.82}{2.87 - 2.82}. \tag{4.3}$$

Solving for Y in (4.3) yields $Y = 2.84$, which is exactly the solution obtained from (4.2). One can see that it may be a bit easier to use interpolation if a table such as Table 4.2 is readily available.

4.2.2   Other Forms of the Grubb Statistic

Farrant (1997) discusses alternative Grubb statistics that essentially serve the same purpose as outlier detectors and are consistent with our aforementioned procedure, but take on a slightly different approach of examining the maximum and minimum sample values and require a different set of critical values to determine significance. The first one we will call Grubb_A and it has the form

$$\text{Grubb}_A = \frac{\max - \min}{S}, \tag{4.4}$$

where $\max$ = maximum value and $\min$ = minimum value and S is the standard deviation of the sample. The purpose of this statistic is to determine if either value, that is, the maximum or minimum value, is a possible outlier. Doing the calculations again from our Method A data, we have

$$\text{Grubb}_A = \frac{284 - 266}{3.995} = 4.51. \tag{4.5}$$

The critical value for this statistic at the 0.05 level from Farrant (1997) is 4.49. See Table 4.3. Obviously, $\text{Grubb}_A > 4.49$. Thus, one of the values (min or max) may be an outlier. We already showed from (4.1) that 266 was a possible outlier candidate.

TABLE 4.3   Critical Values from Grubb_A and Grubb_B Procedures ($\alpha = 0.05$)

n	Grubb_A	Grubb_B	n	Grubb_A	Grubb_B
3	2.00	—	22	4.57	0.498
4	2.43	0.999	23	4.62	0.484
5	2.75	0.982	24	4.66	0.471
6	3.01	0.944	25	4.71	0.459
7	3.22	0..898	26	4.75	0.447
8	3.40	0.852	27	4.79	0.436
9	3.55	0.809	28	4.82	0.425
10	3.68	0.770	29	4.86	0.415
15	4.17	0.618	30	4.89	0.399
20	4.49	0.520	40	5.15	0.328
21	4.53	0.513	50	5.35	0.280

The second alternative we call Grubb$_B$, which is a test again that either the extreme values (min or max) are possible outliers. This formulation takes the form

$$\text{Grubb}_B = 1 - \left(\frac{(N-3)S_{N-2}^2}{(N-1)S^2} \right),$$ (4.6)

where S_{N-2}^2 is the variance of the data with the two extreme values omitted and S^2 is the variance of the full data set. Computing Grubb$_B$ for our Method A data example, we have

$$\text{Grubb}_B = 1 - \left(\frac{(17)\,7.673}{(19)15.958} \right) = 1 - 0.4302 = 0.5698.$$ (4.7)

The critical value for this statistic at the 0.05 level from Farrant (1997) and our Table 4.3 is 0.520. Note that the calculated value of Grubb$_B$ is slightly greater than the critical value of 0.520, leading one to suspect either one or both of the extreme values as possible outliers. As a matter of fact, if one were to repeat the analysis of Table 4.1 with the value 266 omitted and the 19 remaining observations and the value of 284 being the most extreme, the Grubb value would be 2.02. This value would not exceed the value of 2.53, which is the one-sided 0.05 critical value for $n = 19$ in Table 4.2, and it would not exceed the two-sided 0.05 critical value of 2.68. Both critical values for $n = 19$ are derived by interpolation as we did in equation (4.3). Thus, both Grubb$_A$ and Grubb$_B$ are consistent with equation (4.1). Examining Grubb$_A$ and Grubb$_B$, we see that they both are informative in terms of alerting one that either the maximum value or minimum value or both may be possible outliers. One still has to apply equation (4.1) to the suspected data point to determine that it is in fact an outlier.

A note of caution is appropriate here. In all approaches to outlier analysis, one should examine the amount of data that are declared outliers. In general, if 15% or more of the data are considered outliers, then one should examine the assumptions about the data distribution and reevaluate that the data is of sufficient quality and has been sampled and recorded correctly.

4.2.3   Studentized Range Statistic

This statistic is a crude version of the Studentized residual statistic. It is also referred to as the "Extreme Studentized Deviate Test" (ESD). We'll call it the deviate (DEV) test. This is also a very simple statistic such as the Grubb statistic. It is loosely defined as

$$\text{DEV}_i = \frac{(X_i - \text{Mean})}{\text{SD}}, \quad i = 1, 2, \ldots, N,$$ (4.8)

where N = sample size, X_i is the value of the ith observation, $i = 1, \ldots, N$. The value, Mean, is the sample mean and SD is the sample standard deviation. Thus, for our

Method A data in Table 4.1, $N = 20$, Mean $= 2.772$, and SD $= 3.995$. For each observation, we simply compute

$$DEV_1 = \frac{(266 - 277.2)}{3.995} = -2.80$$

$$DEV_2 = \frac{(273 - 277.2)}{3.995} = -1$$

$$\vdots$$

$$DEV_{20} = \frac{(284 - 277.2)}{3.995} = +1.70. \qquad (4.9)$$

We now examine the distribution of these Studentized deviate values. One simple way of approaching this issue is to consider any value outside the range of $(-2.0$ to $+2.0)$ as possible outliers. This is simply a boundary of two or more standard deviations from the mean. The only value satisfying that criteria according to the aforementioned procedure is the cholesterol value associated with $S = -2.80$, or the cholesterol value $= 266.0$, which is consistent with the Grubb procedure. For some studies, the interval $(-3.0, +3.0)$ may be used. The rule is usually for small sample sizes consider the limits $(-2.0, +2.0)$ and for large sample sizes consider the limits $(-3.0, +3.0)$. Small sample sizes may be 35 or less. The advantage of this procedure is that one can consider multiple outliers at one time. All points violating the boundary limits are candidates as outliers. The more common approach is to once again use the values in Table 4.2 as the critical boundary values based on the sample size. One can see once again that for $N = 20$ the critical value for a two-sided 0.05 test or $\alpha = 0.025$ is the value of 2.71 and we consider the absolute value of the computed value of -2.80 or 2.80 as a possible outlier. Thus, it is clear that the ESD test can be thought of as a Grubb test, where the statistic (4.8) is computed for all the points in the distribution and those more extreme than the absolute critical value are considered possible outliers.

4.2.4 Sequential Test of Many Outliers

An alternative method to the Studentized range for testing the possibility of multiple outliers at one time is the test developed by Prescott (1979). This is an extension of the $Grubb_B$ test that we described earlier. It involves the use of the corrected sums of squares. Suppose we have n observations, $Y_1, Y_2, \ldots, Y_n$. We designate the corrected sum of squares of the observed sample as CS^2. This is just the numerator of the variance or

$$CS^2 = \sum_{i=1}^{n} (y_i - \bar{y})^2, \qquad (4.10)$$

where $\bar{y}$ is the usual notation for the sample mean.

The sequential procedure works as follows. Suppose that $y^{(1)}$ is the most extreme observation in the sample. That is to say that $(y^{(1)} - \bar{y})^2$ is the largest squared difference from the mean of all the observations. We then remove $y^{(1)}$ from the sample and compute a new mean

$$\bar{y}_1 = \sum_1 \frac{y_i}{(n-1)}. \tag{4.11}$$

Thus, $\bar{y}_1$ is the mean of the remaining $n-1$ data points once the suspected outlier or extreme value, $y^{(1)}$, has been removed. The corrected sum of squares for these $n-1$ observations is thus defined as

$$CS_1^2 = \sum_1 (y_i - \bar{y}_1)^2. \tag{4.12}$$

We then define the statistic

$$D_1 = \frac{CS_1^2}{CS^2}. \tag{4.13}$$

The value of D_1 is compared to a critical value, which we will discuss. Small values of D_1 are considered significant and thus $y^{(1)}$ would be considered an outlier. Suppose we have k suspect values that we consider possible outliers. We thus perform this process for the k suspect values reducing the sample size each time by one. This will result in k possible statistics $D_1, D_2, \ldots, D_k$, where

$$D_j = \frac{CS_{j-1}^2}{CS_j^2}, \quad CS_j^2 = \sum_j (y_j - \bar{y}_j)^2 \text{ and } \bar{y}_j = \sum_j \frac{y_i}{(n-j)}. \tag{4.14}$$

Note here that following the method by Prescott (1979) that the summations, Σ_j, are evaluated over the $n-j$ observations remaining after the deletion of the suspected extreme observations, $y^{(1)}, y^{(2)}, \ldots, y^{(k)}$, at each step of the sequential analysis for $j = 1, \ldots, k$.

The issue is the number of suspected outliers. That result determines the number of critical values needed for the test of D_j. For example, if there are two suspected outliers and one computes D_1 and D_2, then we have the two critical values, λ_1 and λ_2, each corresponding to D_1 and D_2, respectively, and the rule is that if $D_1 < \lambda_1$, then $y^{(1)}$ is considered outlier. If $D_2 < \lambda_2$, then $y^{(1)}$ and $y^{(2)}$ are both considered outliers. The sequential rule is that only if the last $D_j < \lambda_j$, then all $y^{(j)}$ s are considered outliers. Prescott (1979) gives the critical values for $\alpha = 0.01, 0.025, 0.05,$ and 0.10 for testing up to three possible outliers for sample sizes between 10 and 50. We have provided our own expanded simulated critical values in Table 4.5 for testing up to two possible outliers at $\alpha = 0.05$ for sample sizes from 8 to 55. Our results overlap with Prescott (1979) for the sample sizes he considered.

Let's revisit our Method A data in Table 4.1. The results are given in Table 4.4 for $\alpha = 0.05$. Note that for step $j = 1$ that $D_1 = 0.564 < \lambda_1 = 0.565$ from Table 4.5 ($n = 20$), which indicates that the value 266 is a possible outlier. Note we use the original full data set sample size of 20 to determine the critical values from the table.

TABLE 4.4 Prescott Sequential Test for Outliers ($\alpha = 0.05$) – Method A Data

Step j	Data	Sample Size(n)	Mean	CS_j^2	D_j	λ_j	Decision
—	Full data	20	277.200	303.200	—	—	—
1	Omit 266	19	277.789	171.158	0.564	0.565	Outlier
2	Omit 284	18	277.444	130.444	0.762	0.620	No outlier

TABLE 4.5 Critical Values from Prescott Procedure for Sequential Testing of at Most Two Outliers. ($\alpha = 0.05$)

n	λ_1	λ_2	n	λ_1	λ_2
8	0.222	0.237	22	0.597	0.655
9	0.267	0.288	23	0.610	0.669
10	0.310	0.335	24	0.622	0.681
11	0.345	0.370	25	0.630	0.690
12	0.380	0.420	26	0.643	0.702
13	0.410	0.460	27	0.652	0.710
14	0.440	0.495	28	0.661	0.718
15	0.465	0.520	29	0.669	0.726
16	0.495	0.545	30	0.675	0.735
17	0.515	0.565	35	0.715	0.770
18	0.535	0.585	40	0.745	0.795
19	0.555	0.605	45	0.765	0.810
20	0.565	0.620	50	0.785	0.830
21	0.583	0.642	55	0.808	0.844

Also, for step $j = 2$, we have $D_2 = 0.762 > \lambda_2 = 0.620$ indicating that the value 284 is not an outlier. Also note in Table 4.4 that there is a much greater reduction in the corrected sum of squares from the full data set to step $j = 1$ after removing $y^{(1)} = 266$ than from the step $j = 1$ to step $j = 2$ when $y^{(2)} = 284$ was removed. Thus, one can see that this procedure is certainly consistent with our previous techniques and like the Studentized range has an advantage over the Grubb procedure that multiple outliers can be tested at one time.

4.2.5 Mahalanobis Distance Measure

The next outlier method is called a distance measure, and it is a bit more complicated computationally for detecting outliers. So we'll just examine the plots and interpret the results. The measure is called the "Mahalanobis Distance Measure." The Mahalanobis distance (MD) is a descriptive statistic that provides a relative measure of a data point's distance (like a residual) from a common point. It is a unitless measure and was introduced by Mahalanobis (1936). The MD is used to identify and gauge similarity of points, that is, those that belong to the data set and those that do not.

The simplistic approach is to estimate the standard deviation of the distances of the sample points from the center of mass (perhaps the mean). If the distance between the test point and the center of mass is less than one standard deviation, then we might conclude that it is highly probable that the test point belongs to the set. The further away (perhaps two standard deviations) it is, the more likely that the test point should not be classified as belonging to the set. This intuitive approach can be made quantitative by defining the normalized distance between the test point and the set much like the Studentized distance. By plugging this into the normal distribution, we can derive the probability of the test point belonging to the set. Referring to Figure 4.1, note that one point is above the dashed boundary line. All points below the line are considered part of the data set and any point above the dashed line could be a candidate for an outlier. The one point above the dashed line is observation number 1 of Method A in Table 4.1, or the value 266. This is consistent with our previous techniques utilizing the Grubb statistics as well as the ESD statistic and sequential outlier method.

So one can see that the ESD technique allows us to consider multiple outliers, whereas Grubb just takes one at a time. The MD measure also allows us to consider multiple points as outliers. One can then legitimately ask, "What is the advantage of Mahalanobis over the ESD procedure or the sequential procedure?" The Mahalanobis procedure may actually help one to avoid the problem of masking. This is a good point to introduce the concept of "masking." Barnett and Lewis (1994) define masking as "the tendency for the presence of extreme observations not declared as outliers to mask the discordancy of more extreme observations under investigation as outliers." This statement can be confusing, so we dissect its meaning by example. Suppose we have the following 10 data points from an experiment:

$$7.3, \quad 7.5, \quad 7.7, \quad 8.0, \quad 8.2, \quad 8.3, \quad 8.6, \quad 8.7, \quad 15.7, \quad \text{and} \quad 15.9. \qquad (4.15)$$

Clearly the points 15.7 and 15.9 look very suspect. The mean of the data is 9.59 with a standard deviation of 3.30. Let's apply the ESD procedure and start with the 10th value, 15.9.

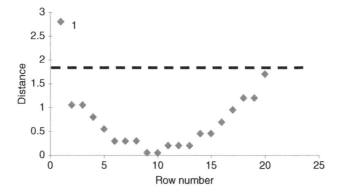

Figure 4.1   Method A Mahalanobis Distances

Thus,

$$DEV_{10} = \frac{(15.9 - 9.59)}{3.30} = 1.91. \tag{4.16}$$

Referring to Table 4.2 for $N = 10$, the critical t-value two-sided $\alpha = 0.05$ is 2.29. Note that $1.91 < 2.29$. Clearly, 15.9 is not considered an outlier. Why is that? What is happening, according to Barnett and Lewis (1994), is that the value 15.9 is being masked by the other extreme value 15.7.

If we eliminate the value 15.7 from the data in (4.15), then we have $N = 9$, mean $= 8.91$, and SD $= 2.66$. The calculation then becomes

$$DEV_{10} = \frac{(15.9 - 8.91)}{2.66} = 2.63. \tag{4.17}$$

Referring to Table 4.2 for $N = 9$ at the 0.025 level and interpolating, we have the critical value of 2.19. Clearly, $2.63 > 2.19$ and the value 15.9 is considered as an outlier. This is an example where the value 15.7 masked the value 15.9 as the outlier. Similarly, if we repeat the aforementioned procedure, we can show that the value 15.9 is masking the value of 15.7 from being declared an outlier. When they are both in the data set, neither is considered an outlier. When either is not in the data set, then the other one is considered an outlier. Now we can consider the advantage of the Mahalanobis distance. Examine Figure 4.2. We see that when the MD procedure is applied, the two extreme points that occur to the far upper right of the plot above the dashed line are in fact the points 15.7 and 15.9 and could be candidates for outliers. Thus, the MD procedure helps get around the issue of masking, and that is the advantage over the Grubb or ESD procedures. It also can detect multiple outliers as seen in Figure 4.2.

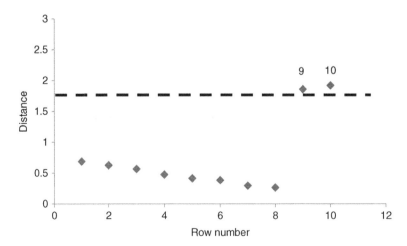

Figure 4.2 Masking Example Mahalanobis Distance

Another term used in outlier analysis is "swamping." Let's again consider the data in (4.15). Block testing is a procedure when one a priori specifies a block or group of observations to be considered as outliers. A simple example will demonstrate block testing's analogue (swamping) to masking. Suppose that, for whatever reason, a researcher wishes to apply a block procedure for outlier detection to the sample in (4.15) and specifies the number of suspicious observations to be the upper three values. That is, the values 8.7, 15.7, and 15.9 are to be tested as being inconsistent with the remainder of the sample against some prespecified underlying probability distribution. A block test applied to these three upper values may well declare them discordant as a unit; the extreme observations 15.7 and 15.9 have overwhelmed the otherwise unexceptional value 8.7. The aforementioned example cites a situation with an obvious upper outlier pair; however, the use of a block test for $k = 3$ upper outliers may declare all three, 15.7, 15.9, and 8.7 as discordant. The nearest neighbor to 15.7 and 15.9 being 8.7 certainly may be a valid member of the population being sampled as was seen by all our aforementioned procedures. Also the MD approach testing for multiple outliers shows clearly in Figure 4.2 that we avoid eliminating valid observations such as 8.7 and clearly helps to avoid either the masking or swamping challenges of the data. For an excellent discussion of swamping and block testing, see Woolley (2002).

We'll see later in the chapter that as the data becomes a bit more complicated that the Mahalanobis distance can continue to be very useful for determining if an observation or group of observations are possible outliers.

4.2.6   Dixon Q-Test for a Single Outlier

The Dixon's Q-test is another outlier test aimed at determining if a single observation is in fact a possible outlier. It is often described in books and articles of Analytical Chemistry in the sections on data handling. This test allows us in a restricted set of a small number of observations (usually 3–10) to examine if one (and only one) observation can be "legitimately" considered an outlier or not. However, Dixon's test has been computed for sample sizes as high as 25 or 30. The calculated statistic is based on Gibbon's (1994) tau statistic, and the algebraic form of the statistic actually depends on the sample size, which we will demonstrate.

The Q-test is based on the statistical distribution of subrange of ratios of ordered statistics or ordered data samples, drawn from a normal population. Thus, a normal (Gaussian) distribution of the data is assumed whenever this test is applied. In the case of the detection and rejection of an outlier, the Q-test cannot be reapplied on the set of the remaining observations as was the Grubb test, or one will experience a loss of power.

The test is very straightforward. The one caution, as stated earlier, is that the actual form of the statistic depends on the sample size, as is seen in the following step 2, which is the only step affected. The rules of steps (1), (3), and (4) remain the same. We proceed as follows:

1. The N values comprising the set of observations under examination are arranged in ascending order:

$$X_1 < X_2 < \cdots < X_N \tag{4.18}$$

2. The statistic that we label as the experimental Q-value (Q_{Exp}) is calculated. This is a ratio defined as the difference of the suspect value from its nearest neighbor divided by the range of the values (Q: rejection quotient). Thus, for testing X_1 or X_N as possible outliers for a sample size of 3–7 we use the following Q_{Exp} statistics:

$$Q_{EXP} = \frac{X_2 - X_1}{X_N - X_1} \qquad Q_{EXP} = \frac{X_N - X_{N-1}}{X_N - X_1}. \tag{4.19}$$

For sample sizes between 8 and 10, the Q_{Exp} statistics for testing X_1 or X_N as possible outliers take the form

$$Q_{EXP} = \frac{X_2 - X_1}{X_N - X_1} \qquad Q_{EXP} = \frac{X_N - X_{N-1}}{X_N - X_2}. \tag{4.20}$$

One sees that the ratio of ranges to compare may change. For the sake of completeness, we list the Q_{Exp} statistics for sample sizes 11–13 and 14–30. For 11–13, we have

$$Q_{EXP} = \frac{X_3 - X_1}{X_{N-1} - X_1} \qquad Q_{EXP} = \frac{X_N - X_{N-2}}{X_N - X_2}. \tag{4.21}$$

For sample sizes between 14 and 30, the Q_{Exp} statistics for testing X_1 or X_N as possible outliers take the form

$$Q_{EXP} = \frac{X_3 - X_1}{X_{N-2} - X_1} \qquad Q_{EXP} = \frac{X_N - X_{N-2}}{X_N - X_3}. \tag{4.22}$$

Since this is such a popular test, we believe it is appropriate to give these options in their entirety.

3. The calculated Q_{Exp} value is compared to a critical Q-value (Q_{Crit}) found in Table 4.6. This critical value should correspond to the significance level, α, at which we have decided to run the test (usually: $\alpha = 0.05$).

4. If $Q_{Exp} > Q_{Crit}$, then the suspected value of X_1 or X_N can be considered as a possible outlier and it possibly can be rejected. If not excluded, then the suspected data point must be retained and used in all subsequent calculations.

The Q-test is a significance test based on the ranges of the data. The null hypothesis associated with the Q-test is this: "There is not a significant difference between the suspect value and the rest of the sample." Any differences must be exclusively attributed to random errors.

Table 4.6 contains a sample of the critical Q-values for our purposes for $\alpha = 0.10$, 0.05, and 0.10 and $N = 3–10$. Expanded tables are published elsewhere.

TABLE 4.6 The Dixon Q-Test

Sample Size	Critical Values for Dixon Q-test		
	$\alpha = 0.10$	$\alpha = 0.05$	$\alpha = 0.01$
3	0.941	0.970	0.994
4	0.765	0.829	0.926
5	0.642	0.710	0.821
6	0.560	0.625	0.740
7	0.507	0.568	0.680
8	0.468	0.526	0.634
9	0.437	0.493	0.598
10	0.412	0.466	0.568

These are standard Q-values and can be found in most references such as Rorabacher (1991) and Bohrer (2010). The values in Table 4.6 are for a two-sided test of significance. For expanded samples sizes up to 30, see Kanji (1993).

Let's take an example. We refer to our bioassay example of Chapter 3 and record 8 reticulocyte counts at log dose 1.699 or dose = 50. The counts are 6.2, 6.3, 6.3, 6.4, 6.4, 6.5, 6.6, and 7.4. The suspected outlier is 7.4. Since our sample is of size 8, we use equation (4.20), or the equation,

$$Q_{EXP} = \frac{X_N - X_{N-1}}{X_N - X_2},$$

where $X_N = 7.4$, $X_{N-1} = 6.6$, and $X_2 = 6.3$. Therefore,

$$Q_{EXP} = \frac{7.4 - 6.6}{7.4 - 6.3} = \frac{0.8}{1.1} = 0.727. \tag{4.23}$$

Thus, for $N = 8$ in Table 4.6 at $\alpha = 0.05$, we have $Q_{Crit} = 0.526$ and thus $Q_{Exp} > Q_{Crit}$ or $0.727 > 0.526$. We then reject 7.4 at the level of $\alpha = 0.05$ as not being significantly different from the remainder of the sample and it is, thus, considered a possible outlier. Examining Table 4.6 further, note that $0.727 > Q_{Crit} = 0.634$ and if the initial stated $\alpha = 0.01$, then we could consider the value, 7.4, to be a possibly significant outlier at the 0.01 level.

We have been assuming thus far that our data has been following an underlying normal distribution. Although Dixon's test involves an assumption of normality, it is effective both at detecting outliers and is robust to departures from normality. Chernick (1982) showed that when the sample size is 3–5, the test retains its significance level when the distribution is not normal, specifically, uniform, exponential, and so on. Because it is based on ratios of the spacings between order statistics, Chernick (1982) believed that it should be robust and also believes the robustness property would hold up in larger samples as well, but this has not yet been fully investigated. Also the following methodology of the box plot depends on the interquartile range (IQR). A data set need not be normal to utilize the properties of the IQR. Thus, the box plot methodology can be applied to data not following a normal pattern.

4.2.7 The Box Plot

A box plot is a way of summarizing data measured on an interval or continuous scale and is used often in many applications for exploratory data analysis. It is a graph used to show the shape of a distribution. The elements of the graph are its central value, its spread, and the minimum and maximum values of the data. Any data point or points that lie beyond these extreme values are treated as outliers. The box plot rule has been applied to detect outliers for univariate and multivariate data in medical data and clinical chemistry data (Solberg and Lahti 2005). As an example, let us consider the following 13 reticulocyte counts from an earlier experiment: 5.0, 6.0, 6.3, 6.3, 6.4, 6.4, 6.5, 6.6, 7.0, 7.1, 7.2, 7.3, and 8.5. Figure 4.3 is the box plot of the data. The upper and lower boundaries of the box define the IQR or the values of the 25th and 75th percentiles, which are 6.3 and 7.15, respectively. The line within the box is the median at value = 6.5. The horizontal line passing through the box is the mean at value = 6.65. The horizontal lines below and above the box are the minimum and maximum values of the data without the outliers or 6.0 and 7.3, respectively. The lower and upper extreme points are the outlier points, 5.0 and 8.5, respectively.

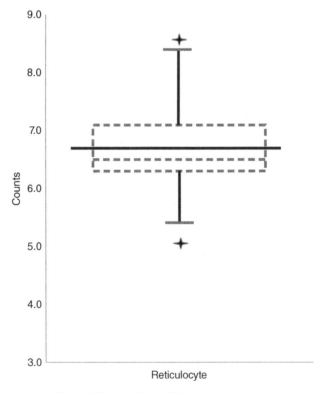

Figure 4.3 Box Plot of Reticulocyte Counts

TABLE 4.7 Nonparametric Reticulocyte Outlier Data

Counts	Median	Absolute Deviation from the Median	Median of Absolute Deviations	Test Statistic	Possible Outlier?
6.5	6.5	0.0	0.5	0.0	No
6.4	6.5	0.1	0.5	0.2	No
6.4	6.5	0.1	0.5	0.2	No
6.6	6.5	0.1	0.5	0.2	No
6.3	6.5	0.2	0.5	0.4	No
6.3	6.5	0.2	0.5	0.4	No
6.0	6.5	0.5	0.5	1.0	No
7.0	6.5	0.5	0.5	1.0	No
7.1	6.5	0.6	0.5	1.2	No
7.2	6.5	0.7	0.5	1.4	No
7.3	6.5	0.8	0.5	1.6	No
5.0	6.5	1.5	0.5	3.0	Yes
8.5	6.5	2.0	0.5	4.0	Yes

Thus, the box identifies the middle 50% of the data (25th–75th percentiles), the median and the extreme values. The box plot assumes the data does not deviate significantly from normality. Also note that the box plot can capture more than one suspected outlier at a time. If a graph is not available, a rule of thumb is to compute the IQR, that is, $7.15-6.3 = 0.85$. One then takes 1.5 times the IQR or $1.5 \times 0.85 = 1.27$. One then takes the first quartile minus the value 1.27 or $6.3 - 1.27 = 5.03$ and if any points lie below this value, then it is considered an outlier candidate. On the other end, take the third quartile plus this value or $7.15 + 1.27 = 8.42$ and if any points lie above this value it is considered an outlier candidate. One can see that the points 5.0 and 8.5 are certainly candidates. Oftentimes if the data is normal, the multiplier is taken as 1.35 instead of 1.5. In either case, 5.0 and 8.5 would be considered possible outliers.

4.2.8 Median Absolute Deviation

Although we've been discussing robust methods of outlier detection, one may still wish to examine a methodology in which the data is free of any known underlying distribution. These methods of outlier detection are known as nonparametric or distribution-free approaches. One such test is called the Median Absolute Deviation (MAD) test. Note Table 4.7 and the reticulocyte counts that we have been considering.
   We proceed as follows:

1. Compute the median of the count data, which is the first column of Table 4.7. This median value is 6.5, as seen repeated in the second column of the table.
2. Compute the absolute value of deviations of original input data from this median value of 6.5. This is the column titled "Absolute Deviation from the Median."

3. Compute the median of these absolute deviations, which is 0.5 as seen repeated in the Table in the column "Median of Absolute Deviations."

4. Compute the ratio of the absolute deviation from step 2 and median 0.5 from step 3 for each row in the table. This is the column titled "Test Statistic."

5. If this ratio is greater than an assigned critical value, then consider the value as a possible outlier.

The critical value is really arbitrary. Some suggestions for the critical value by Miller (1991) are 2.5, which is considered moderately conservative, 3.0, very conservative, or 2.0, poorly conservative. If we use 2.5, then from the last two columns of Table 4.7, we see that the original counts of 5.0 and 8.5 are possible outliers.

4.3 COMBINED METHOD COMPARISON OUTLIER ANALYSIS

4.3.1 Further Outlier Considerations

In this section, we want to combine what we've learned about method comparisons discussed in Chapter 3 and outlier considerations. Suppose we have information on Method B, that is, cholesterol readings from the same individual's blood sample, but measured with an alternative instrument, B, as seen in Table 4.1. Now let's plot the results of Method B versus Method A. This is seen in Figure 4.4. From the plot, one often thinks that we can get an idea of the agreement between the two approaches. Another method is to compute the correlation between the two methods. This is not appropriate, as the correlation tells of similarity of measures in a certain direction, but not agreement in magnitude.

We will examine these concepts. First examine the plot of Method A versus Method B. Note from Table 4.1 (asterisk, first column) and Figure 4.4 that the

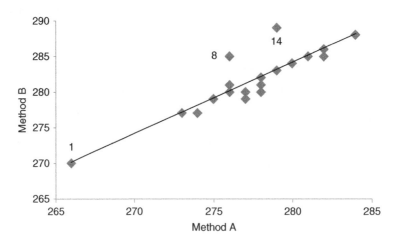

Figure 4.4   Method A versus Method B Total Cholesterol

agreement of cholesterol readings from the two methods is discrepant by a factor 10 for subject numbers 14 and 9 for subject number 8. All the rest differ by a factor of 4 or 5. Also, all the B readings are greater than the A readings. Thus visually, it looks pretty good, $R^2 = 0.81$, which is a correlation of 0.90, but there certainly are some discrepancies. Observations 8 and 14 would be considered outlier candidates as well since their residuals (distance from the line) are 4.8 and 5.8, respectively, outside the limits of $(-2.0 \text{ and} + 2.0)$. Observation number 1 on Figure 4.4 appears to be separated from the other data points, which are mostly to the upper right, and one may have a tendency to label that point as a possible outlier. However, not so fast. It deserves further discussion. Let's look at the components of the line. The equation of the line is

$$\text{METHOD\_B} = 4.0171504 + 1.0006596 \ \text{METHOD\_A}. \qquad (4.24)$$

Our line here has a significant slope (slope $= 1.0006, p = 0.0001$), and the intercept is not significantly different from 0 (intercept $= 4.0172, p = 0.9018$). As a matter of fact from Chapter 3, if we were to test that the slope $= 1$, the p-value for this particular test question is $p = 0.498$, or the slope does not differ statistically from the value 1. Thus, the line fit indicates pretty good agreement between Methods A and B. We are not ready to declare that is the case. We'll come back to that issue. For now, let's stay with the outlier discussion. Staying with the line, let's consider an added line examining technique and put a 95% density ellipse over the scatter plot. This is the region where we expect in repeated samplings of Method A versus Method B that 95% of the data points will occur. This is seen in Figure 4.5. Note that the ellipse covers most points except for observations 1, 8, and 14. Observations 8 and 14 are outside the ellipse and fairly distant from the line. Observation 1 is outside the ellipse, but right on the line. Note that from Table 4.1 the Method A, B coordinates for the points 1,

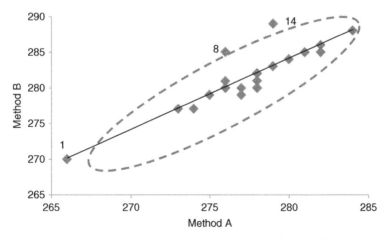

Figure 4.5 Method A versus Method B with 95% Density Ellipse

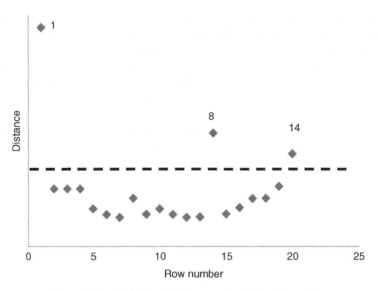

Figure 4.6   Mahalanobis Distance for Methods A and B

8, and 14 are (266, 270), (276,285), and (279,289), respectively. Note how close the observation 1 coordinates are to each other compared to the other two observations. This is why one has to examine the situation completely before jumping to the conclusion that a point is an outlier. Clearly, observation 1 is a legitimate data point. It is just a low-cholesterol reading that is attested by both methods. Observations 8 and 14 could be possible outliers.

Let's look at the Mahalanobis distance measure (MD) we discussed earlier. This is seen in Figure 4.6. Here we have all three data points above the line containing diamonds, indicating a possible outlier. The "row number" title on the horizontal axis in the plot is merely the observation number. Here, as with the density ellipse in Figure 4.5, not knowing the true nature of the experiment, one can easily think of observation 1 as a possible outlier, which it is not.

So from this example, we see that one has to be careful how one defines an "outlier."

Do they cause large residuals in the linear plot indicating discrepancy between the Method A and B readings (observations 8 and 14), or are they merely far outside the major cluster of points (higher cholesterol readings) that make up most of the data? They may be legitimate points as observation 1, but slightly out of the range of the norm, and still valid for measuring agreement.

4.3.2   Combined Method Comparison Outlier Analysis – Refined Method Comparisons Using Bland – Altman

We are going to revisit the Bland–Altman procedure discussed in Chapter 3. We generate the Bland–Altman plot for measure of agreement between Methods A and B

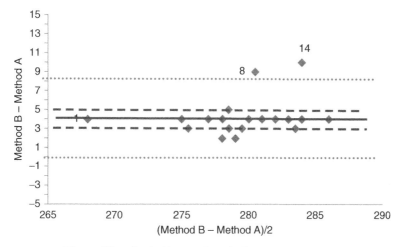

Figure 4.7 Bland–Altman of Method A versus Method B

from Table 4.1. This is seen in Figure 4.7. Let's do some calculations needed to assess the amount of agreement.

The mean difference, $d = 4.2$. This is the solid line in the center of Figure 4.7. Note immediately that this line is not near the value, 0.The upper and lower dashed lines are the 95% confidence interval on the mean (3.281, 5.119). The standard deviation of the difference $S = 1.963$. We examine the limits, that is, 2 standard deviations below the mean and 2 standard deviations above the mean:

$$d - 2S = 4.2 - 2(1.963) = 0.274$$

$$d + 2S = 4.2 + 2(1.963) = 8.126 \quad\quad (4.25)$$

Thus, $(d - 2S, d + 2S) = (0.274, 8.126)$. Note that the quantity, $d \pm 2S$, is sometimes referred to as the bias. The mindset is often that if the computed differences (Method B–Method A) fall within the interval, $(d - 2S, d + 2S)$, then we can use the two methods interchangeably. The limits of this interval are the two dotted lines on Figure 4.7. Note that all the differences of Method B minus Method A derived from Table 4.1, except observation 8, which has the value = 9, and observation 14, which is the value = 10, do in fact fall within the limits. The width of the interval $(d - 2S, d + 2S) = (0.274, 8.126)$ is rather wide and does not cover the value 0. According to Bland–Altman, this would indicate poor agreement between the two methods. More relevant to the topic of this chapter, it appears that the observations 8 and 14 may be outliers. We confirm that further. As in Chapter 3, Bland–Altman allow us to take yet another closer look at the limits $(d - 2S, d + 2S)$. They put a 95% confidence interval on the limits, $d - 2S$ and $d + 2S$, which we construct as follows.

Recall that the standard error (SE) of $d - 2S$ as well as $d + 2S$ is $(3S^2/N)^{1/2}$. In our sample, $N = 20$ and $S = 1.963$. Thus, SE $= (3 * 3.85/20)^{1/2} = 0.7599$.

The t-value for a 95% confidence interval on $N - 1 = 19$ degrees of freedom (df) = 2.093. So the 95% confidence interval on $d - 2S$ is

$$d - 2S - 2.093(0.7599) \text{ and } d - 2S + 2.093(0.7599)$$

or

$$(0.274 - 1.5905, \ 0.274 + 1.5905) = (-1.316, \ 1.864). \tag{4.26}$$

Similarly, the 95% confidence interval for $d + 2S$ is

$$d + 2S - 2.093(0.7599) \text{ and } d + 2S + 2.0938(0.7599) \tag{4.27}$$

or

$$(8.126 - 1.5905, \ 8.126 + 1.5905) = (6.536, \ 9.716).$$

Note that the upper limit of $d + 2S$ is 9.716 and this upper limit is violated by observation 14, which has the value of Method B–Method A = 10. Thus, this observation definitely appears to be an outlier, as it is outside the 95% region of the upper two standard deviation limit of the mean difference, d.

Thus, for several reasons it appears that there is <u>not</u> good agreement between Methods A and B. They are as follows:

- The p-value for the differences of the two methods is $p = 0.0001$ by simple paired t-test.
- The limits $(d - 2S, \ d + 2S) = (0.274, 8.126)$ are rather wide.
- Two observations (8 and 14) fall outside the limits of $d - 2S$ and $d + 2S$.
- One observation (14) falls outside the upper 95% confidence limit of $d + 2S$, which may indicate an outlier.

Thus, we have a two-for-one deal, where we are using the Bland–Altman method of agreement to also label possible outliers.

Let's look at another example. Note that in Table 4.1, we also have Method C to compare to Method A. It looks as if these values are very close to each other over the 20 patient total cholesterol values (TC) that were run through the lab. Note that some values of one method are slightly greater than the values of the other method and vice versa. The equation of this line is

$$\text{METHOD\_C} = -0.1387 + 1.0039 \ \text{METHOD\_A}.$$

Note that for the linear fit on Figure 4.8 that all the values of Method C regressed on Method A are well within the 95% density ellipse, except subject 1, which is a lower value, but is right on the line. The residuals are all within $+2$ or -2 standard

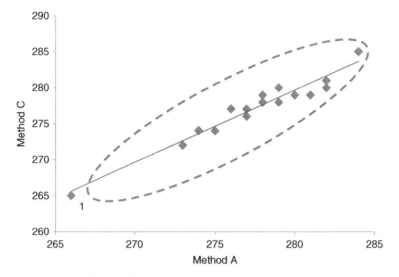

Figure 4.8 Method C Regressed on Method A

errors of 0 (the predicted line) indicating no outliers. Here $R^2 = 0.938$ or correlation, $R = 0.968$. The slope is significantly different from zero, $p = 0.0001$ and the line goes through the origin, $p = 0.9349$. Also, for a test of the slope $= 1.0$ the p-value $= 0.4744$, which indicates the slope not significantly deviating from 1.0. Thus, just based on linearity, Methods A and C appear to be in agreement. Let's examine the Mahalanobis Distance (MD) plot seen in Figure 4.9 for Methods A and C. The only discrepant point is observation $= 1$, which is a naturally low-cholesterol value and does not speak to the agreement issue. We take it one step further and examine the Bland–Altman plot of Figure 4.10.

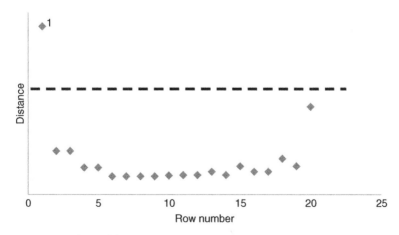

Figure 4.9 Mahalanobis Distance of Methods A and C

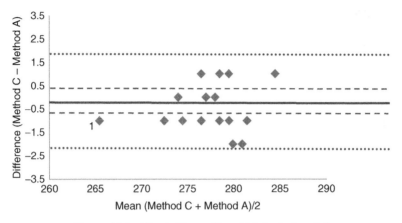

Figure 4.10 Bland–Altman Plot of Methods A and C

From the data in Table 4.1 and Figure 4.10, the mean difference $= -0.3$ and standard deviation, $S = 1.0311$. We use the formulae already provided to compute $(d - 2S, d + 2S) = (-2.362, 1.762)$, whose boundary is seen by the dashed line on Figure 4.10. Note all points are within the dashed lines or within two standard deviations of the mean difference, even observation 1 which one could mistakenly label as an outlier based on the MD plot. The 95% confidence interval on $d - 2S$ is $(-3.198, -1.526)$ and the 95% confidence interval on $d + 2S$ is $(0.926, 2.598)$. Note that no observed differences are below -3.198 or above 2.598.

Thus, for several reasons it appears that there is good agreement between Methods A and C. They are as follows:

- The p-value for the differences of the two methods is $p = 0.2088$; no difference based on the paired t-test.
- The limits $(d - 2S, d + 2S) = (-2.362, 1.762)$ are rather narrow, indicating perhaps good agreement.
- No values fall outside the limits of $d - 2S$ and $d + 2S$.
- No values fall outside the lower 95% confidence limit of $d - 2S$ or outside the upper limit of $d + 2S$.

Thus, this appears to be a situation in which we have good agreement and no outliers to concern us.

All these methods that we presented in this chapter are good methods to get a statistical interpretation of the outlier severity. However, as we stated earlier, one must truly understand the physical situation to determine if an observation or group of observations are in fact outliers or just a suspected value that may in fact be a legitimate data point, as we've seen with observation number 1, which is an individual with a comparatively low cholesterol value.

4.4　SOME CONSEQUENCES OF OUTLIER REMOVAL

We have been discussing methodologies for detecting outliers and have cautiously considered whether or not particular values may in fact be considered as outliers. Osborne and Overbay (2004) give an excellent general outline of outlier detection and consequences of removal with a rather comprehensive bibliography on the subject. We have stated that one must really be convinced that an outlier is in fact a legitimate outlier and then deal with it accordingly. Obviously, it has to be explained, and we've covered this consideration in our introduction to this chapter. As Osborne and Overbay (2004) point out, conceptually, there may be strong arguments for removal or alteration of outliers. The analyses reported in their paper empirically demonstrate the benefits of outlier removal. In their examples, both correlations and t-tests appeared to show significant changes in statistics as a function of removal of outliers, and in most of their presentation they contend that the accuracy of estimates was enhanced. Also, in most situations, errors of inference were significantly reduced, a prime argument for screening and removal of outliers. We are going to show here by demonstration and use of the techniques presented earlier how inferences can change when one does remove outliers. We'll keep our example simple and demonstrate with only one outlier. Consider the data in Table 4.8. We revisit a reticulocyte count example. The goal is to compare the two groups, A and B, and determine if they differ statistically with respect to reticulocyte count. There are 13 observations in Group A and 11 observations in Group B. This is from our laboratory example we discussed earlier. In Chapter 2, we learned how we can compare the groups, either by a simple independent sample t-test or a nonparametric test. Not germane to our goal here is the fact that Group A does not deviate significantly from normality, but

TABLE 4.8　Comparative Reticulocyte Example

Reticulocyte Counts	
Group A	Group B
7.5	6.9
6.7	7.0
6.9	6.8
7.0	7.3
7.3	7.3
7.2	6.3
7.4	7.0
7.2	7.0
7.3	7.1
7.5	7.2
7.8	7.3
7.3	—
9.5	—

Group B does. We'll do a nonparametric comparison of the two populations based on our data.

Using Grubb's test, the value of 9.5 in the last row of the Group A data may be considered an outlier. The mean of Group A is 7.431, with a standard deviation of 0.682. Thus, the Grubb value is

$$\frac{(9.5 - 7.431)}{0.682} = 3.035. \tag{4.28}$$

Interpolating from Table 4.2, the Grubb critical value for $N = 13$ and $\alpha = 0.01$ is 2.59. Clearly $3.035 > 2.59$, which obviously indicates that the value, 9.5, may be considered an outlier.

Assuming that it is an outlier, if we left 9.5 in the data set and computed a nonparametric Wilcoxon test to compare Groups A and B, the p-value $= 0.0403$. If we remove it from the data set, the p-value $= 0.0659$. Clearly the disposition of the value, 9.5, has consequences for declaring the groups as statistically different (supposed outlier remains in the data set) or not statistically different (9.5 removed from the data set). This simple example demonstrates the sensitivity and consequences of one's decision and the necessity for making the correct assessment when it comes to handling of suspected outliers.

4.5   CONSIDERING OUTLIER VARIANCE

4.5.1   The Cochran C test

When data come from different laboratories we may wish to conduct interlaboratory comparisons. The methods we have been discussing can be used, for example, to determine if a mean or median from one laboratory is an outlier compared to the means or medians from the other laboratories. The Grubb statistics, Studentized range, Dixon, and so on can be used as we have demonstrated. Another consideration is that one may believe that the variance or dispersion from a particular laboratory or sampled batch is rather large compared to the other laboratories. We do have at our disposal in much of the available statistical software the ability to test the homogeneity of variances or equality of variances among many groups. These well-known tests are the Bartlett test, Browne–Forsyth, and Levene test among others. This is important information. However, we take this step a bit further to determine if a particular variance is an outlier variance compared to others. The Cochran test (often called the Cochran C test) can be used in this case for testing of a suspected outlying variance. It is a very simple test in which the suspected variance is compared with the sum of all the variances. Suppose we have L laboratories and we run n_i replicate experiments for each laboratory, $i = 1, \ldots, L$. Let S_i be the standard deviation of the runs within each laboratory. Suppose one suspects the value, S, from one of the laboratories to be an outlier. The Cochran statistic for this test takes the simple form

$$C_G = \frac{S^2}{\sum\limits_{i=1}^{L} S_i^2}, \qquad (4.29)$$

where $G = \sum_{i=1}^{L} n_i/L$ or the average number of replicates per group. The restricting assumption here is that the n_i per group or laboratory are nearly equal or within ± 1 of each other. There are other limitations that we will point out later. The test consists of comparing the calculated value, C_G, to the table of critical values for the Cochran test found in Kanji (1993), Konieczka and Namiesnik (2009), ISO Standard 5725-2 (1994), and van Belle et al. (2004). These tables are generally limited to the α level of 0.05 and 0.01 and generally do not have all values of interest for L and G. One has to derive these omitted values through a process of interpolation, which we shall consider later. We'll demonstrate by an example. Suppose that we have $L = 10$ laboratories, each with $G = 10$ replicate laboratory values of interest with standard deviations, which we put in ascending order, that is, 0.12, 0.137, 0.198, 0.237, 0.256, 0.305, 0.416, 0.451, 0.502, 0.712. Our standard deviation of concern, say from Lab 10, is $S = 0.712$. See Figure 4.11 and note the larger bar for that lab. Thus,

$$C_G = \frac{S^2}{\sum\limits_{i=1}^{L} S_i^2} = \frac{0.712^2}{1.42251} = \frac{0.5069}{1.42251} = 0.3564. \qquad (4.30)$$

Accordingly, the 0.05 level critical value for $L = 10$ and $G = 10$ is 0.2439. See our Table 4.9 for which we have computed the 0.05 critical values for values of L ranging from 2 to 15 and G ranging from 2 to 20. Some published tables give L from 2 to 180 and G from 2 to 150 or higher. Note at $L = 10$ and $G = 10$, we have the critical

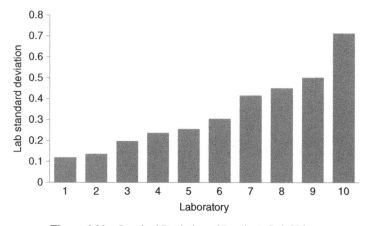

Figure 4.11   Standard Deviation of Replicate Lab Values

TABLE 4.9   Sample of Critical Values for Cochran C Outlier Variance Test ($\alpha = 0.05$)

$L =$ Number of Labs	\multicolumn{9}{c}{$G =$ Number of Replicates per Laboratory}

$L =$ Number of Labs	2	4	6	8	10	12	14	16	18	20
2	0.9985	0.9392	0.8772	0.8332	0.8010	0.7765	0.7570	0.7417	0.7223	0.7164
3	0.9669	0.7977	0.7070	0.6531	0.6167	0.5902	0.5698	0.5543	0.5348	0.5289
4	0.9065	0.6839	0.5894	0.5365	0.5018	0.4769	0.4579	0.4436	0.4258	0.4205
5	0.8413	0.5981	0.5063	0.4564	0.4241	0.4012	0.3839	0.3710	0.3548	0.3500
6	0.7807	0.5321	0.4447	0.3980	0.3682	0.3471	0.3313	0.3195	0.3048	0.3004
7	0.7270	0.4800	0.3972	0.3536	0.3259	0.3064	0.2918	0.2810	0.2675	0.2635
8	0.6798	0.4377	0.3594	0.3185	0.2927	0.2746	0.2611	0.2511	0.2387	0.2350
9	0.6385	0.4027	0.3285	0.2901	0.2659	0.2491	0.2365	0.2272	0.2157	0.2122
10	0.6020	0.3733	0.3028	0.2666	0.2439	0.2281	0.2163	0.2076	0.1968	0.1963
11	0.5697	0.3482	0.2811	0.2468	0.2254	0.2105	0.1994	0.1913	0.1812	0.1782
12	0.5410	0.3264	0.2624	0.2299	0.2096	0.1955	0.1851	0.1774	0.1678	0.1650
13	0.5152	0.3074	0.2463	0.2152	0.1960	0.1827	0.1728	0.1655	0.1565	0.1538
14	0.4919	0.2907	0.2321	0.2025	0.1841	0.1714	0.1620	0.1551	0.1466	0.1441
15	0.4709	0.2758	0.2195	0.1912	0.1737	0.1616	0.1526	0.1460	0.1379	0.1355

value of 0.2439. Clearly 0.3564 > 0.2439 and the variance value $= (0.712)^2$ would be declared an outlier variance, which would perhaps dictate examining that laboratory's procedures.

It is important to note here that the published tables contain gaps, and one can derive these critical values by interpolation or simulation. In our particular case of Table 4.9, note that the critical values for $G = 15$ are not listed. Suppose we had $L = 5$ labs and $G = 15$ replicates per laboratory. Note in the table that we have $G = 14$ and $G = 16$ for $L = 5$, which yields the critical values 0.3839 and 0.3710, respectively. Obviously, our critical value of interest, X, is somewhere between 0.3839 and 0.3710. We repeat the interpolation technique as we did for equation (4.3), that is, we have the simple setup:

G	Critical Value
14	0.3839
15	X
16	0.3710

To solve for X, we use interpolation or the simple formula:

$$\frac{15 - 14}{16 - 14} = \frac{X - 0.3839}{0.3710 - 0.3839}. \tag{4.31}$$

Mathematically, all we do is perform the same arithmetic operation on either column of the above table, set them equal to each other and solve for X. This is interpolation.

This yields the solution for X as 0.3775, which is the critical value at $\alpha = 0.05$ for $L = 5$ and $G = 15$. We'll see other examples of interpolation throughout the book.

For the EXCEL statistical user, this test, called the Cochran C test, can be run very easily in Microsoft XLSTAT-Pro by clicking on Menu, navigating to Outlier Analysis and choosing the Cochran C test. The output includes a summary of the input data as well as the exact C_G as we calculated here and the exact p-value, which in this case will obviously be below 0.05. Other outlier analyses we discussed such as the Grubb test and Studentized range test can also be performed in the EXCEL XLSTAT Outlier platform.

4.5.2   Cochran G Test

There are limitations to the Cochran C test as noted by 't Lam (2010):

1. The C test gives equal weight to all standard deviations, which is OK for balanced designs (i.e., equal number of replications per group), but not for unbalanced designs, which should allow greater weight to those groups having a larger sample size.

2. The tables for the C test are available for determining critical values for testing if a variance is exceptionally large, but not if one is exceptionally small. This limits all tests to one-sided upper tail tests. One can clearly miss if a low S_i is an outlier.

3. As noted earlier, tables are limited to $\alpha = 0.01$ and $\alpha = 0.05$ and for some published tables 't Lam (2010) has noted that there are transcription errors, which the author has corrected in his publication. We have made appropriate corrections to our tables as well.

Clearly, one would want to test for lower and upper variance outliers. This is precisely the method of 't Lam (2010), which we will demonstrate here. The theoretical development is quite complex. Thus, naturally we will give the essentials of the practical computations of the test, which is called the Cochran G test and demonstrate the procedure with an example.

As earlier, let S_i be the standard deviation of the runs within each lab, $i = 1, \ldots, L$. The test statistic for lab $j, j = 1, \ldots, L$ is defined as

$$G_j = \frac{v_j S_j^2}{\displaystyle\sum_{i=1}^{L} v_i S_i^2}, \tag{4.32}$$

where $v_i = n_i - 1$, n_i being the number of replicates for lab i and S_j is the standard deviation of interest. Note that this is a weighting factor for each laboratory or group. We are going to test that G_j meets certain critical upper and lower values based on the F-statistic, which we discussed in Chapter 2. The procedure for testing each laboratory as a possible outlier takes the form of a two-sided interval,

$$G_{\text{LL}}(\alpha, v_j, v_{\text{SUM}}, L) < G_j < G_{\text{UL}}(\alpha, v_j, v_{\text{SUM}}, L). \tag{4.33}$$

Here $v_{\text{SUM}} = \sum_{i=1}^{L} v_i$ and α of course is the alpha level of interest. G_{LL} and G_{UL} are the lower and upper limits of this equality, which we will define in detail. The goal is to determine if G_j lies within the interval defined by (4.33). The decision is simply if G_j lies within the interval, then the laboratory value S_j is not an outlier. If G_j lies outside the interval, then S_j is an outlier. Before proceeding, let's look at some data. The data setup is in Table 4.10. We have six laboratories and in the third column we have, n_i, or the number of experiments or replicates in lab $i = 1, \ldots, 6$, which ranges from 2 to 8. Note the standard deviations in the S_i column range from 0.013 for laboratory 1 to 4.623 for laboratory 6. Intuitively, it looks as if these two extreme values may be candidates for outlier standard deviations. Once we compute G_j for each laboratory, $j = 1, \ldots, 6$, from (4.32), we then compute G_{LL} and G_{UL}. These values look complex but are quite easy to derive with some computing help. We define the lower bound as

$$G_{\text{LL}}(\alpha, v_j, v_{\text{SUM}}, L) = \left[1 + \frac{(v_{\text{sum}}/v_j) - 1}{F_c((1 - \alpha/2L),\ v_j, v_{\text{sum}} - v_j)} \right]^{-1}, \tag{4.34}$$

where $F_c((1 - \alpha/2L), v_j, v_{\text{sum}} - v_j)$ is the critical value of the F-distribution with probability $1 - \alpha/2L$, where L, of course, in this case $= 6$. If $\alpha = 0.05$, then $1 - \alpha/2L = 1 - 0.05/12 = 0.99583$. For example, for the first laboratory value, the numerator degrees of freedom (df) is $v_1 = 2$ and the denominator df is $v_{\text{sum}} - v_1 = 23 - 2 = 21$. Thus, if $j = 1$ or the first laboratory, then $F_c((1 - \alpha/2L), v_j, v_{\text{sum}} - v_j) = F(0.99583, 2, 21) = 0.004179$. This value is easily calculated from EXCEL XLSTAT-Pro. The authors were able to derive this F-critical value easily from the F inverse function using EXCEL. Thus, the lower limit or G_{LL} from (4.34) is

$$G_{\text{LL}}(0.05, 2, 23, 6) = \left[1 + \frac{(23/2) - 1}{0.004179} \right]^{-1} = 0.000398 = 0.0004, \tag{4.35}$$

which is the G_{LL} value seen in Table 4.10 for the first laboratory.

We define the upper bound as

$$G_{\text{UL}}(\alpha, v_j, v_{\text{SUM}}, L) = \left[1 + \frac{(v_{\text{sum}}/v_j) - 1}{F_c((\alpha/2L),\ v_j, v_{\text{sum}} - v_j)} \right]^{-1}, \tag{4.36}$$

where $F_c((\alpha/2L), v_j, v_{\text{sum}} - v_j)$ is the critical value of the F-distribution with probability $\alpha/2L$, where L of course in this case $= 6$. Again, if $\alpha = 0.05$, then $\alpha/2L = 0.05/12 = 0.00417$. For example, for the first laboratory value, the numerator degrees of freedom (df) is $v_1 = 2$ and the denominator df is $v_{\text{sum}} - v_1 = 23 - 2 = 21$. Thus, if $j = 1$ or the first lab, then $F_c((\alpha/2L), v_j, v_{sum} - v_j) = F(0.00417, 2, 21) = 7.1960$. This value is again easily calculated from EXCEL XLSTAT Pro. Thus, the upper limit or G_{UL} from (4.36) is

$$G_{UL}(0.05, 2, 23, 6) = \left[1 + \frac{(23/2) - 1}{7.1960}\right]^{-1} = 0.4070, \qquad (4.37)$$

which is the G_{UL} value seen in Table 4.10 for the first lab. From (4.32), the value of

$$G_1 = \frac{v_1 S_1^2}{\sum\limits_{i=1}^{L} v_i S_i^2} = \frac{0.000338}{182.07397} = 1.9 \times 10^{-6}. \qquad (4.38)$$

Note from our aforementioned calculations that G_1 does not lie in the interval $(G_{LL}, G_{UL}) = (0.0004, 0.4070)$. Thus, the variance or standard deviation for laboratory 1 does not pass the test noted in (4.33). Therefore, it is labeled a "No" in the Pass column of Table 4.10. To establish a more rigorous criterion or statistical decision as to whether a standard deviation is an outlier based on a p-value from an F ratio 't Lam (2010) defined the F-statistic for each laboratory as

$$F_j = \frac{[(v_{sum}/v_j) - 1]}{(G_j^{-1} - 1)}. \qquad (4.39)$$

Thus the p-value in usual statistical terms is the probability of achieving an F ratio greater than F_j or using the notation of 't Lam (2010), the p-value is defined as

$$\gamma_j = 1 - \text{Probability}(F \text{ ratio} < F_j) \qquad (4.40)$$

TABLE 4.10 Cochran G Test Results for the Six Laboratories

Cycle	Lab(i)	n_i	v_i	S_i	G_{LL}	G_j	G_{UL}	Pass	γ_i	δ_i
1	1	3	2	0.013	0.0004	1.9×10^{-6}	0.4070	No	0.9999	0.00002
	2	4	3	1.496	0.0031	0.0369	0.4761	Yes	0.8567	0.1433
	3	2	1	1.635	0.0001	0.0147	0.3171	Yes	0.5727	0.4273
	4	5	4	2.002	0.0094	0.0881	0.5356	Yes	0.7651	0.2394
	5	8	7	2.015	0.0495	0.1561	0.6808	Yes	0.8740	0.1259
	6	7	6	4.623	0.0329	0.7043	0.6367	No	0.0009	0.0009
2	2	4	3	1.496	0.0039	0.0369	0.5011	Yes	0.8745	0.1255
	3	2	1	1.635	0.0001	0.0147	0.3321	Yes	0.5912	0.4088
	4	5	4	2.002	0.1145	0.0881	0.5639	Yes	0.7987	0.2013
	5	8	7	2.015	0.0586	0.1561	0.7155	Yes	0.9050	0.0950
	6	7	6	4.623	0.0392	0.7043	0.6698	No	0.0024	0.0024
3	2	4	3	1.496	0.0067	0.1247	0.6296	Yes	0.6454	0.3546
	3	2	1	1.635	0.0001	0.0496	0.4245	Yes	0.4068	0.4068
	4	5	4	2.002	0.0192	0.2977	0.7019	Yes	0.3775	0.3775
	5	8	7	2.015	0.0981	0.5279	0.8626	Yes	0.3665	0.3665

with numerator df $= v_j$ and denominator df $= v_{sum} - v_j$. Again, this probability is easily calculated in EXCEL XLSTAT and is seen in the next to the last column in Table 4.10. 't Lam (2010) pointed out that when conducting a two-sided G test as we are, it is convenient to use a decision parameter that is either equal to γ_j or $1 - \gamma_j$, whichever value is the smallest. Thus, we define

$$\delta_j = \min(\gamma_j, 1 - \gamma_j). \qquad (4.41)$$

This is the p-value associated with the hypotheses

$$H_0 : S_j \text{ is not an outlier versus}$$

$$H_1 : S_j \text{ is an outlier.}$$

The value δ_j is seen in the last column of Table 4.10 calculated for all the $S_j, j = 1, \ldots, 6$ and thus we see that we would in fact reject the null hypothesis that S_1 is not an outlier since $\delta_j = 0.00002 < 0.05$ and is certainly less than the value $\alpha/2L = 0.00417$, which is relevant in a multiple comparison situation as we discussed in Chapter 2. The aforementioned procedures (4.32–4.34), (4.36), (4.39–4.41) are repeated for laboratories 2–6, which completes cycle 1 of Table 4.10. We note that both values of G_1 and G_6 fall outside the range of (G_{LL}, G_{UL}) and each has a δ_j value ($j = 1, 6$) that is quite small. The rule is to eliminate the S value for the smallest δ_j and repeat the process with that laboratory excluded. Note when such is done for cycle 2 (excluding lab 1) then laboratory 6 is eliminated as S_6 is now considered an outlier by our aforementioned procedures. There are no more outliers noted when a third cycle is attempted as seen in Table 4.10. Thus, we conclude for this example that S_1 and S_6 are considered outlier standard deviations. One can see that the G test, unlike the C test, allows one to test for unusually small as well as unusually large outlier variances.

REFERENCES

Barnett V and Lewis T. (1994). Outliers in Statistical Data, 3rd ed., John Wiley & Sons, New York, NY.

Böhrer A. (2010). One-sided and two-sided critical values for Dixon's outlier test for sample sizes up to $n = 30$. Economic Quality Control 23(1): 5–13.

Chernick MR. (1982). A note on the robustness of Dixon's ratio test in small samples. American Statistician 36(2): 140.

Farrant TJ. (1997). Practical Statistics for the Analytical Scientists. A Bench Guide, Royal Society of Chemistry, Cambridge (ISBN 0-85404-442-6).

FDA (2006). Guidance for Industry Investigating Out-of-Specification (2006). Test Results for Pharmaceutical Production. FDA. CDER.

Gibbons, RD (1984). Statistical Methods for Groundwater Monitoring, John Wiley & Sons, New York, NY.

Iglewicz B and Hoaglin DC. (1993). How to Detect and Handle Outliers, American Society for Quality Control, Milwaukee, WI.

International Organization for Standardization (1994). Accuracy (trueness and precision) of Measurement Methods and Results – Part 2: basic Method for Determination of Reproducibility of a Standard Measurement Method. [ISO Standard 5725-2, 1994] Geneva, Switzerland.

Kanji GK. (1993). 100 Statistical Tests. Sage Publications Ltd., London.

Konieczka P and Namiesnik J. (2009). Quality Assurance and Quality Control in the Analytical Chemical Laboratory – A Practical Approach. CRC Press, Boca Raton, FL.

Mahalanobis PC. (1936). On the generalised distance in statistics. Proceedings of the National Institute of Sciences of India; 2(1): 49–55.

Miller J. (1991). Reaction time analysis with outlier exclusion: bias varies with sample size. Quarterly Journal of Experimental Psychology A 43(4): 907–912, 10.1080/14640749108400962.

Osborne JW and Overbay A. (2004). The Power of outliers (and why researchers should always check for them). Practical Assessment, Research and Evaluation; 9(6). Retrieved July 28, 2013 from http://PAREonline.net/getvn.asp?v=9&n=6 .

Prescott P. (1979). Critical values for a sequential test for many outliers. Applied Statistics 28: 36–39.

Rorabacher DB. (1991). Statistical treatment for rejection of deviant values: critical values of Dixon Q parameter and related subrange ratios at the 95 percent confidence level. Analytical Chemistry 63(2): 139–146.

Solberg HE and Lahti A. (2005). Detection of outliers in reference distributions: performance of horn's algorithm. Clinical Chemistry 51(12): 2326–2332.

't Lam, RU. (2010). Scrutiny of variance results for outliers: Cochran's test optimized. Analytica Chimica Acta 659: 68–84.

van Belle G, Fisher LD, Heagerty PJ and Lumley TS. (2004). Biostatistics: A Methodology for the Health Sciences, 2nd ed., Wiley Interscience, Hoboken, NJ.

Wooley T. (2002). The effect of swamping on outlier detection in normal samples. Joint Statistical Meetings. American Statistical Association, ENAR, WNAR, August 11-15, New York, NY. www.amstat.org/sections/srms/Proceedings/y2002/files/JSM2002-000362.pdf.

5

STATISTICAL PROCESS CONTROL

5.1 INTRODUCTION

So far, we have been using statistical design and analysis that allows us to examine changes in inputs (e.g., concentrations) and to observe corresponding changes in outputs (e.g., response to the analyte measured). Conversely, there are statistical graphical techniques that we can use to test the homogeneity (constancy of variance) of a system, stability of the laboratory process, and maybe spot possible outliers. This all involves statistical process control (SPC). We are unable in just one chapter to give a full course in SPC but can certainly talk about the relevant concepts and applications in a laboratory setting. We start with a brief summary of what SPC is all about.

We refer to a process, for example, such as repeat laboratory experiments for a specified period of time. SPC is the method used to monitor processes, track conformity to specifications, and evaluate the actual measurement process. SPC enables a laboratory monitor to track, recognize, and possibly reduce process variability and improve the process by using statistical analytic tools such as control charts. Laboratories often use alternative terminology by referring to the use of SPC methodology in their internal quality control program as statistical quality control (SQC).

Introduction to Statistical Analysis of Laboratory Data, First Edition.
Alfred A. Bartolucci, Karan P. Singh, and Sejong Bae.
© 2016 John Wiley & Sons, Inc. Published 2016 by John Wiley & Sons, Inc.

The control chart that we will investigate in detail in the following section is the most used and most successful SPC tool. It was developed by Walter Shewhart (1931, 1939) of the Bell Laboratories in the early 1920s. The control chart helps you record data with certain specific elements incorporated into the chart and lets you see when an aberration or unusual event occurs, usually a very high or low observation in light of a "typical" or "expected" process performance. The statistical motivation of control charts is an attempt to distinguish between two sources of process variation. Dr. Shewhart referred to these as "chance" variation or that which is intrinsic to the process and will always be present and "assignable or uncontrolled" variation, which stems from external sources and indicates that the process is out of statistical control. Deming (1950, 1952) introduced SPC to Japan after the Second World War and relabeled "chance" variation as "common cause" variation and "uncontrolled" as "special cause." Nonetheless, the underlying meanings remain, as with Shewhart's original definitions. To be more specific, examples of common cause variation, which Deming also called "random," can be a poorly trained worker or lab technician, fluctuations in the instrument or machinery, and in manufacturing normal variation of raw materials. Examples of special cause variation, sometimes leading to biased results, are seasonal effects in environmental sampling and instrument calibration being off. See Chapter 7 for further discussion of bias. These can all lead to what is referred to as an out-of-control result or event. Various tests, such as the outlier techniques discussed in Chapter 4 and methods discussed in this chapter, can help determine when an out-of-control event has occurred. Also, one must be aware of the distinction between the intent of Shewhart's original work and what followed, such as Deming's purpose and intent. Shewhart's goal can be described as that of helping to get the process into that "satisfactory state" which one might then be content to monitor. The mathematical sophistication that followed, such as the work of Deming (1975), was in that context and for the purpose of process improvement, as opposed to only process monitoring. A true follower to Deming's principles would probably adhere to the philosophy and aim of continuous improvement in the process. This is very much true today, where in many cases SPC continually evolves with the aim of refinement of the process, as we learn how to better reign in the uncontrollable or special causes of variation.

There are certain procedures or steps in conducting an SPC analysis. One first wishes to identify the cause of variation in order to address it and remedy it. This is not always obvious or straightforward because of complex manufacturing or laboratory operations involving many interrelated variables. Statistical Control Charts, which we will discuss, help to distinguish between common causes and special causes of variation. Once these sources of variation are determined, we wish to remove them, correct them, or certainly help stabilize the system. Some procedures of correction may involve recalibrating the instrument, establishing storage standards to minimize deterioration of raw materials, and possibly others. In Chapter 8, we discuss further statistical techniques for identifying significant causes of variation in a system or experiment. Once a process is free of special causes of variation, it is said to be stable although there still may be variation due to random causes.

The next step is to estimate the process capability or the ability of the process to meet a particular statistical standard beyond the control chart, which we will demonstrate. Once these conditions are met, one would wish to establish and carry out a plan to monitor, improve, and assure the quality of the process, for example, establish charting procedures, maintenance, training, and record keeping, in order to keep variation in check.

Our goal in this chapter is to describe the major components of SPC and explain their application to interpreting the statistical behavior of a process.

5.2 CONTROL CHARTS

Control charts help us to detect special causes of variation and measure or monitor common causes of variation. One must keep in mind that the purpose of the control chart is not to make precise probability statements, but it is a guide as to when investigative action may be needed. We may be inclined to make adjustments to our procedures once the data layout in the control chart is examined. We set up and explain the elements of the control chart by example.

Suppose we refer to Chapter 3, Section 3.2.2, where we examined accuracy of a dilution series of bacterial control RNA (DAP) ranging from 5 to 40 attomoles, which was added to buffer to get a standard. Recall, we then measured percent recovery of mRNA from the spiked sample. The full procedure is explained in Section 3.2.2. We conduct this procedure six times on each of 7 days for a total sample size of 42. Thus, we have a batch of six samples on each day, and we measure the percent recovery from each of the samples. The data is seen in Table 5.1 and in Figure 5.1. This figure is, of course, the actual percent recovery for the six experiments or samples plotted against the day it was performed for each of the 7 days, day 1 to day 7. If this were an industrial setting, it would be like having seven lots and six samples per lot. Examining Figure 5.1, it appears that the samples look fairly stable or comparable over the 7 days. As a matter of fact, if we do a one-way ANOVA testing the equality of the means, the overall p-value is 0.3979. This gives us a very simple picture of the behavior of the process over time. Let's take another look at the process behavior data in Figure 5.2. Here we have an individual measurement chart that connects the percent recovery points within and across the 7 days. Again, the pattern looks fairly random and fairly well in control in that there are not any obvious points that deviate from a fairly regular pattern. There are sophisticated ways of putting upper and lower control limits or bounds to check the process using the ranges within the samples. However, given the mean and standard deviation of the entire process, which are 100.07 and 0.840, respectively, one often initially checks for the stability of the process to see if the data points are within 3 standard deviations of the mean of the distribution by examining the upper and lower 3 standard deviation boundaries, which are 102.59 and 97.55. One sees from Figures 5.1 and 5.2 that all the points appear to be within these limits. A note of caution here is that some texts and publications will use 3 standard deviations or 3 sigma units as a guide and claim that this encloses 99.0% of the distribution. Such is not quite the case. For normal samples, 99% of the distribution is

TABLE 5.1   Percent Recovery Data for MRNA data

Percent Recovery Spike/Standard ($N = 42$)

Day	Batch Size (Number of Samples per Day)	Percent Recovery	Mean	Standard Deviation	Median	Range
1	6	99.8, 101, 100.9 101.2, 100.4, 99.1	100.40	0.812	100.65	2.1
2	6	100.1, 99.6, 98.8 100.8, 101.3, 99.6	100.03	0.905	99.85	2.5
3	6	98.9, 99.2, 99.1 101.6, 100.1, 99.2	99.68	1.026	99.20	2.7
4	6	99.8, 101, 100.9 101.2, 100.4, 99.1	100.40	0.812	100.65	2.1
5	6	99.3, 98.9, 99.5 99.8, 100.3, 99.5	99.55	0.472	99.50	1.4
6	6	100.1, 99.6, 98.8 100.8, 101.3, 99.6	100.03	0.905	99.85	2.5
7	6	99.8, 101, 100.9 101.2, 100.4, 99.1	100.40	0.812	100.65	2.1

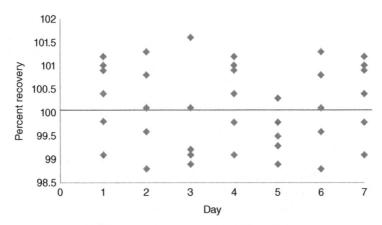

Figure 5.1   Percent Recovery Spike/Standard

within 2.57 standard deviations of the mean. Three standard deviations enclose about 99.9% of the distribution. Some control chart software will use the 3 sigma limits or 3 standard deviations. Some software will, in fact, construct the control limits based on 99.0% and use the correct normal deviate of 2.57. It is important for the reader to know this subtle distinction and feel comfortable with the method being used.

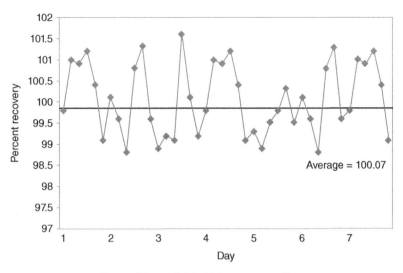

Figure 5.2 Individual Measurement Chart

5.2.1 Means (*X*-bar) Control Charts

From this data, we will construct a control chart for the means and the ranges to examine a range of limits on these values. We start with the means. Figure 5.3 is the control chart often called the *X*-bar ($\overline{\text{x}}$) chart of the means. We explain the components of this chart.

1. The seven points on the chart are the means of the percent recovery or average of the six samples for each of the 7 days. Here, the sample size within each day is $n = 6$, or we have the same number of samples per day. Such is not always the case, and this will affect the formulation of the limits as we will see later.

2. The solid center line is the overall mean from the 42 data points, which we denote by $\overline{x}$. This has the value 100.071.

3. The dashed line at the top is the overall mean plus k (normal deviate) standard deviations (sometimes called the k sigma limit) $\overline{x} + (k\hat{\sigma}/\sqrt{n})$ or the upper limit of the x values, called the upper control limit (UCL). Here, $k = 2.57$ from the standard normal curve to denote the 99% critical value and $n = 6$ for the six experiments per day. The value, $\hat{\sigma}$ is the process standard deviation, which is a bit involved, and we discuss this later. Thus, in this context,

$$\text{UCL} = \overline{x} + \frac{k\hat{\sigma}}{\sqrt{n}} = 101.135. \tag{5.1}$$

4. The dashed line at the bottom is $\overline{x} - (k\hat{\sigma}/\sqrt{n})$ or the lower limit of x values, also called the lower control limit (LCL). This value is 99.008. So in this situation, we have

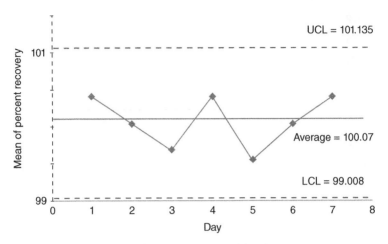

Figure 5.3   Control Chart of the Means

$$LCL = \bar{x} - \frac{k\hat{\sigma}}{\sqrt{n}} = 99.008. \tag{5.2}$$

The calculation of the estimate of the process standard deviation, $\hat{\sigma}$, is governed by the sample size of each of the batches or days in order to reduce bias. It is also a function of the sample ranges, R_1, R_2, ... , R_7 for each day 1 to day 7, respectively. This is seen in the explanation and tables presented in Burr (1976) and AIAG (2005). Here we give the results. Thus, the value of $\hat{\sigma}$ is given by

$$\hat{\sigma} = \frac{\dfrac{R_1}{d_{2(n_1)}} + \cdots + \dfrac{R_7}{d_{2(n_7)}}}{N}, \tag{5.3}$$

where N = number of samples = 7, and as per Burr (1976), since $n_i = 6$, which is the size of all our subsamples for $i = 1, 2, ... , 7$, then $d_{2(n_i)} = 2.534 d_{2(n_i)}$ is defined as the expected value of the range of n independently normally distributed variables with unit standard deviation. A range of values for $d_{2(n_i)}$ for various values of n_i are given in Table 5.3. Note in our example, as already stated, $n_i = 6$ for all i. So in Table 5.3, $d_{2(n_i)} = 2.534$. Also, the values of the ranges, $R_1, R_2, ... , R_7$, from Table 5.1 are, respectively, $2.1, 2.5, 2.7, 2.1, 1.4, 2.9$, and 2.1. Thus, in equations (5.1) and (5.2), $\hat{\sigma} = 1.012$. We let $k = 2.57$ for 99% limits on our data and we arrive at (UCL, LCL) = (101.135, 99.008) for Figure 5.3.

After calculating the UCL and LCL, one sees if any points of Table 5.1 or Figure 5.2 exceed these limits. Sometimes, one takes the approach that these "outlier" points should not be included in constructing the mean chart limits as these points may be due to special causes (not common) and will skew the chart. They often discard these "out-of-control" points and calculate new values for $\bar{x}$, s, UCL, and LCL. One then checks the data to ensure that all data lies within the new UCL and LCL. If not, then repeat the discarding and recalculating of data until all points are within UCL and LCL.

If one repeatedly cannot draw a control chart with all data within the limits, it indicates that the process is out of control, plagued with special causes of variation, and one cannot utilize a control chart until these special causes are eliminated.

Recall, because we used the UCL and LCL as $(\bar{x} \pm 3s)$, that only one result per 100 should lie beyond the control limits, that is, >99% of all data lies within $\bar{x} \pm 3s_x$ if random variation is the cause. Such a process is then said to be "in control." If data exceeds the control limits more frequently, this is due to special causes of variation and the process (or data set) is said to be "out of control."

Now, let's suppose we follow the aforementioned rule (which is not always wise, as useful and legitimate data may be eliminated) and go back to Table 5.1 and eliminate all points above the UCL of 101.135 and below the LCL of 99.008 in Figure 5.2. This involves the elimination of 10 data points or about one quarter of the data. Thus, we are left with 32 data points (Table 5.2). The new control chart is seen in Figure 5.4. Note the difference in the form of the UCL and LCL of Figure 5.4 as opposed to Figure 5.3. In Figure 5.3, all the sample sizes within day were the same, that is, $n_i = 6$ for $i = 1, 2, \ldots, 7$. However, in Figure 5.4 the samples are of size 4 or 5. The limits that are most narrow are from the samples of size 5, and the limits that are the widest are from samples of size 4. As one would expect, the smaller the sample size, the lesser the precision. The formulae for the center line or overall

TABLE 5.2 Reduced Percent Recovery Sample from Table 5.1

		Percent Recovery Spike/Standard ($N = 32$)				
Day	Batch Size (Number of Samples per Day)	Percent Recovery	Mean	Standard Deviation	Median	Range
1	5	99.8, 101, 100.9 100.4, 99.1	100.24	0.796	100.4	1.9
2	4	100.1, 99.6 100.8, 99.6	100.02	0.568	99.85	1.2
3	4	99.2, 99.1 100.1, 99.2	99.40	0.469	99.20	1.0
4	5	99.8, 101, 100.9 100.4, 99.1	100.24	0.796	100.40	1.9
5	5	99.3, 99.5, 99.8 100.3, 99.5	99.68	0.390	99.50	1.0
6	4	100.1, 99.6 100.8, 99.6	100.02	0.568	99.85	1.2
7	5	99.8, 101, 100.9 100.4, 99.1	100.24	0.796	99.45	1.9

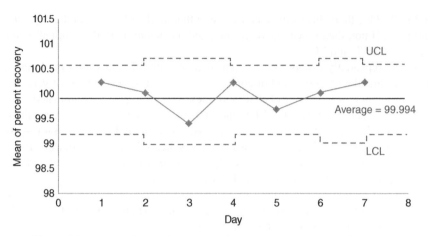

Figure 5.4   Control Chart of the Means for the Reduced Data Set of Table 5.2

mean, $\overline{X}$, is now the weighted average of the means across the 7 days. The UCL and LCL, respectively, for each of the days have the form

$$\overline{X} \pm \frac{k\hat{\sigma}}{\sqrt{n_i}}, \quad i = 1, 2, \dots, 7. \tag{5.4}$$

Here, $k = 3$ for 3 sigma limits or the 99% limit. Sometimes, one uses $k = 2$ for 2 sigma limits or the 95% limits on the process. The overall estimate of the process standard deviation, $\hat{\sigma}$, has the form

$$\hat{\sigma} = \frac{\frac{s_1}{c_{4(n_1)}} + \cdots + \frac{s_7}{c_{4(n_7)}}}{N}, \tag{5.5}$$

where n_i and s_i are the sample size and standard deviation of the ith group, respectively,

$$i = 1, 2, \dots, 7.$$

$N =$ number of batches or days for which $n_i \geq 2$. Here, $N = 7$.

$c_{4(n_i)}, i = 1, 2, \dots, 7$, is the expected value of the standard deviation of n_i independent normally distributed variables with unit standard deviation.

The quantity $c_{4(n_i)}$ is a small sample bias correction for the standard deviation. The subscript of 4 is a standard notation for this particular value found in most texts discussing the construction of the control chart with unequal batch (sample) sizes across days, lots, and so on. The computation of this value is a bit involved and often done computationally. However, there are references that will give the exact value of c_4 based on the sample size. For example, see Burr (1976) and AIAG (2005). Also see Table 5.3 for a list of these constants for a range of sample sizes. If one wishes to compute $\hat{\sigma}$ in (5.5), the statistics, n_i and s_i, as well as the constants,

TABLE 5.3 Sample of Control Chart Constants

Subgroup Size (n_i)	A_2	d_2	c_4	d_4	c_5
2	1.880	1.128	0.7979	3.267	0.6028
3	1.023	1.693	0.8862	2.574	0.4633
4	0.729	2.059	0.9213	2.282	0.3889
5	0.577	2.326	0.9400	2.114	0.3412
6	0.483	2.534	0.9515	2.004	0.3076
7	0.419	2.704	0.9594	1.924	0.2820
8	0.373	2.847	0.9650	1.864	0.2622
9	0.337	2.970	0.9693	1.816	0.2459
10	0.308	3.078	0.9727	1.777	0.2321
11	0.285	3.173	0.9754	1.744	0.2206
12	0.266	3.258	0.9776	1.717	0.2107
13	0.249	3.336	0.9794	1.693	0.2020
14	0.235	3.407	0.9810	1.672	0.1943
15	0.223	3.472	0.9823	1.653	0.1874
20	0.180	3.735	0.9869	1.585	0.1613
25	0.153	3.931	0.9896	1.541	0.1438

Control Chart Constants

TABLE 5.4 Sigma Constants for Equation (5.5)

Sigma (σ) Statistics

i	n_i	s_i	$c_{4(ni)}$
1	5	0.7956	0.9400
2	4	0.5679	0.9213
3	4	0.4690	0.9213
4	5	0.7956	0.9400
5	5	0.3898	0.9400
6	4	0.5679	0.9213
7	5	0.7956	0.9400

$c_{4(n_i)}, i = 1, 2, \ldots, 7$, from Table 5.3 are given in Table 5.4. The value, $\hat{\sigma} = 0.6709$ in Figure 5.4, where the group sample size varies according to Table 5.2. Thus, in Figure 5.4, we have the calculations of the UCL and LCL from equation (5.4). Therefore, for $n_i = 4$ the UCL = 100.99 and the LCL = 99.00. For $n_i = 5$ the UCL = 100.90 and the LCL = 99.20. The tighter control limits obviously are aligned with $n_i = 5$. Examining Table 5.2, one sees that all values are above 99.0 and three values of the percent recovery are values of 101, which are very close to 100.99. It may be reasonable here to say that this reduced size process is in fact in control.

5.2.2 Range Control Charts

We are going to return to the original data in Table 5.1 and consider the control chart for the range of the values, which we call the R chart. Much of the computation that we already performed applies here as well. The advantages of the range chart are that range charts can be used when one cannot construct an individual measurement chart such as Figure 5.2 or a mean (i.e., $\overline{X}$ chart) such as Figure 5.3. For example, such may be the case when standards are not available or too unstable to retain and analyze repeatedly. This situation may not be that common in the laboratory setting. Nonetheless, the actual unknown samples are usually split and analyzed in replicate and the range ($X_{largest} - X_{smallest}$), that is, R values, their mean, and the UCL for the range are calculated and plotted on an R chart. The disadvantage of the R charts is that they do not necessarily detect bias (special cause variation) because this is canceled out when the difference between analyses is calculated, leaving only random (common) causes of variation. The range chart, however, does show the precision of the values. The range chart for our original percent recovery data in Table 5.1 is seen in Figure 5.5. We construct the range chart as follows:

Suppose we have n ranges, R_i, $i = 1, 2, \ldots, n$.

1. Calculate the arithmetic average of the ranges $\left(\overline{R} = \dfrac{\sum R_i}{n} \right)$ and draw it as a solid horizontal line.

2. The UCL for the range, which we label as UCL_R, can be calculated as $\overline{R} \pm 3s_R$, where s_R is the standard deviation of the ranges, but more often it is calculated using a formula derived from the normal distribution, that is, $UCL_R = d_4 \cdot \overline{R}$, where d_4 is a factor obtained from tables used in constructing control charts.

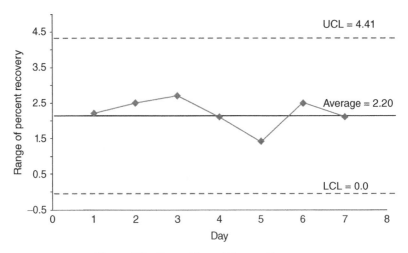

Figure 5.5   Range Chart of Percent Recovery

They also depend on the subsample size, and examples of values of d_4 are seen in Table 5.3. The UCL_R is drawn as a dotted line on R charts.

3. The LCL is often not used for range charts. Instead, the zero line serves as the LCL.

Referring to Table 5.1 and Figure 5.5, the ranges for each of the 7 days, as we've seen earlier, are 2.1, 2.5, 2.7, 2.1, 1.4, 2.5, and 2.1, respectively. Their average is 2.20 as seen as the middle line in Figure 5.5. To compute the UCL or $UCL_R = d_4 \cdot \overline{R}$ defined earlier, the value of d_4 from Burr (1976) or AIAG (2005) and from Table 5.3 for $n = 6$ is 2.004. Thus, the $UCL_R = 2.004\,(2.20) = 4.41$ as seen as the UCL in Figure 5.5. As seen in this figure, it appears that the ranges are well within the control limits.

We now refer back to the reduced sample in Table 5.2 and determine how to construct a range control chart when the sample sizes in each of the groups are not the same. The chart appears in Figure 5.6. Note that the UCL appears in the same format as Figure 5.4 for the mean control chart. However, here the center line, or average, changes as well as the UCL_R. That is to say, these statistics depend on the sample size. The center line for each day is determined by the formula

$$\overline{R}_i = d_{2(n_i)}\hat{\sigma}, \qquad (5.6)$$

where $d_{2(n_i)}$ is defined earlier and $\hat{\sigma}$ is defined by equation (5.3), and in this case $\hat{\sigma} = 0.6470$. The ranges, R_i, from Table 5.2 for the samples $i = 1$ to 7 are 1.9, 1.2, 1.0, 1.9, 1.0, 1.2, and 1.9, respectively. In this case, $d_{2(n_i)} = 2.059$ for $n_i = 4$ and $d_{2(n_i)} = 2.326$ for $n_i = 5$, $i = 1, 2, \dots, 7$. (Table 5.3). This yields the values of the

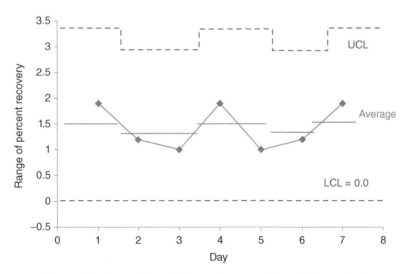

Figure 5.6 Range Chart of Percent Recovery (Unequal Group Size)

center line of 1.33 for $n_i = 4$, $i = 2, 3, 6$, and 1.50 for $n_i = 5$, $i = 1, 4, 5, 7$. The UCL, UCL_R, is computed as

$$\text{UCL}_R = d_{4(n_i)}\overline{R}_i, \tag{5.7}$$

where $d_{4(n_i)}$ is the expected range of a normally distributed sample of size n_i, $i = 1, 2, \ldots, 7$.

In this case, $d_{4(n_i)} = 2.282$ for $n_i = 4$, $i = 2, 3, 6$ and $d_{4(n_i)} = 2.114$ for $n_i = 5$, $i = 1, 4, 5, 7$. This yields the values of the $\text{UCL}_R = 2.82$ for $n_i = 4$ and 3.42 for $n_i = 5$. (Table 5.3). See Figure 5.6 for the range chart of these differing sample size batches. One can see that for the plotted ranges, the process looks much in control, as the plot is well within the upper and lower control limits.

5.2.3 The S-Chart

Thus far, we have been discussing the mechanics behind construction of the X-bar chart, or mean chart, as well as the range chart. It may be of interest to decide if the variation of the process is out of control or could be corrected. We discussed the detection of variation as a possible outlier in Chapter 4. Our goal here is to look at it from a slightly different perspective, that is to say, in the process control perspective. We'll refer to the standard deviation control chart as the S-chart. We will refer to the data in Table 5.1 in which we have all groups of the same size. In this case, the center line of the S-chart is merely the average of the seven standard deviations (Standard Deviation column in Table 5.1) or

$$\overline{S} = \sum_{i=1}^{7} \frac{S_i}{7}, \tag{5.8}$$

where S_i, $i = 1, 2, \ldots, 7$, are the seven standard deviations from Table 5.1, that is, $S_1 = 0.812$, $S_2 = 0.905$, $\ldots, S_7 = 0.812$. Thus, $\overline{S} = 0.8206$. The UCLs and LCLs are defined as

$$\text{UCL} = c_{4(n_i)}\hat{\sigma} + kc_{5(n_i)}\hat{\sigma} \tag{5.9}$$
$$\text{LCL} = \max(c_{4(n_i)}\hat{\sigma} - kc_{5(n_i)}\hat{\sigma}, 0).$$

In this case, $n_i = 6$ for all $i = 1, 2, \ldots, 7$. Thus, referring to Table 5.3, we have $c_{4(n_i)} = c_{4(6)} = 0.9515$. $\hat{\sigma}$ is defined as in equation (5.5). The value $N = 7$ is used in equation (5.5). Thus, solving equation (5.5) for these values we have $\hat{\sigma} = 0.8624$. For the S-chart, we let $k = 3$ or 3 sigma values distance from the center line or $\overline{S}$. We define another constant, $c_{5(n_i)}$. This value is the standard error of the standard deviation of n_i independent observations from a normal population with unit standard deviation. For a given sample size, these constants are found in Table 5.3. Again, as for all constants found in Table 5.3, this constant is found in expanded sample size tables in most texts dealing with SPC. Note for case $n_i = 6$ for all $i = 1, 2, \ldots, 7$ we have $c_{5(6)} = 0.3076$.

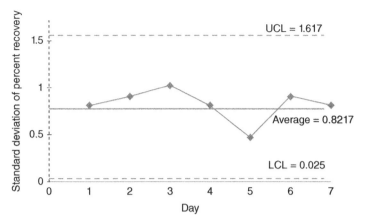

Figure 5.7   *S*-Chart for Percent Recovery (Equal Group Sizes)

Thus, given the values, $c_{4(6)} = 0.9515$, $c_{5(6)} = 0.3076$, $k = 3$ and $\hat{\sigma} = 0.8624$ from equation (5.9), we have

$$UCL = (0.9515)(0.8624) + 3(0.3076)(0.8624) = 1.617. \qquad (5.10)$$

Note that $c_{4(n_i)}\hat{\sigma} - kc_{5(n_i)}\hat{\sigma}, 0 = (0.9515)(0.8624) - 3(0.3076)(0.8624) = 0.02475$. Thus, from (5.9), we have

$$LCL = \max(0.025, \ 0) = 0.025, \qquad (5.11)$$

which is seen in Figure 5.7. Clearly, the process standard deviations look pretty much in control as all standard deviations in Table 5.1 are between the UCL and LCL.

For the sake of completion, let's demonstrate the *S*-chart construction for the case in which we have unequal observations per group or day as we have in Table 5.2. One way to construct the center line for the new *S*-chart is to take the weighted average of the S_i's in Table 5.4, which is 0.6344, and then multiply this value by $c_{4(n_i)}$ in the last column of Table 5.4. This gives a different value of the constant, $c_{4(n_i)}$ (i.e., 0.9213 for $n_i = 4$ and 0.9400 for $n_i = 5$), as well as the center line for the changing sample size (4 or 5) for each day. Doing this simple calculation gives the center line value for days 1, 4, 5, and 7 as 0.596. This line for days 2, 3, and 6 has the value 0.584. These values are very close and may not be that distinguishable graphically in Figure 5.8. For computing the UCL and LCL, the constant $c_{5(n_i)}$ also changes value for sample sizes 4 and 5. That is to say $c_{5(4)} = 0.3889$ and $c_{5(5)} = 0.3412$. Also, recall, the value of $\hat{\sigma}$, which depends on $c_{4(n_i)}$ alone in equation (5.5), will have a new value with unequal sample sizes. This new value is $\hat{\sigma} = 0.6707$. So for days with sample size 4, we have

$$UCL = (0.9213)(0.6707) + 3(0.3889)(0.6707) = 1.400. \qquad (5.12)$$

For sample 5, we have

$$UCL = (0.9400)(0.6707) + 3(0.3412)(0.6707) = 1.317. \qquad (5.13)$$

The LCL for both sample sizes of 4 and 5 is equal to 0. To see this note that for $n_i = 4$, we have

$$c_{4(4)}\widehat{\sigma} = 0.9213(0.6707) = 0.6179, \qquad (5.14)$$

$$kc_{5(4)}\widehat{\sigma} = 3(0.3889)(0.6707) = 0.7825.$$

For $n_i = 4$, we have

$$c_{4(5)}\widehat{\sigma} = 0.9400(0.6707) = 0.6305, \qquad (5.15)$$

$$kc_{5(5)}\widehat{\sigma} = 3(0.3412.6707) = 0.6865.$$

Note in both cases,

$$\max(c_{4(n_i)}\widehat{\sigma} - kc_{5(n_i)}\widehat{\sigma}, 0) = 0. \qquad (5.16)$$

Thus, in Figure 5.8 it appears that the process standard deviations are well in control.

5.2.4 The Median Chart

The median control chart is a simple alternative to the means chart in the case that the data may not be normally distributed. Again, let's first take the case where the group sample sizes are equal, as in Table 5.1. The center line of the median chart is the average of the 7 medians or MED = 100.05. The upper and lower control limits are defined:

$$UCL = MED + A_2\overline{R}, \qquad (5.17)$$

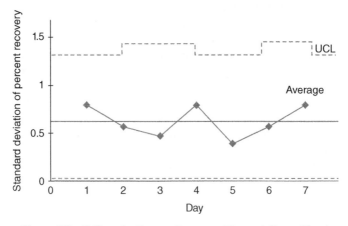

Figure 5.8 *S*-Chart for Percent Recovery (Unequal Group Sizes)

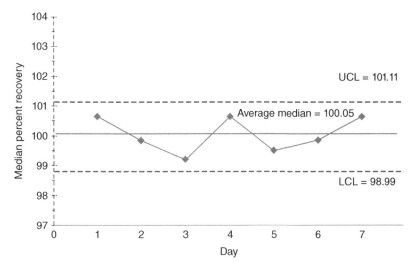

Figure 5.9   Median Control Chart for Percent Recovery (Equal Group Sizes)

$$LCL = MED - A_2\overline{R},$$

where A_2 is a constant depending on the group size and according to Table 5.3 for $n_i = 6$ for all $i = 1, 2, \ldots, 7$, we have $A_2 = 0.483$. $\overline{R}$ is the average of the ranges in Table 5.1 which is the value 2.20. Thus,

$$UCL = 100.05 + 0.483(2.20) = 101.11 \tag{5.18}$$
$$LCL = 100.05 - 0.483(2.20) = 98.99.$$

Based on the medians, the process appears to be in control (Figure 5.9).

Now suppose once again we are dealing with unequal group or subgroup sizes as in Table 5.2. Much like the control chart for means, the center line can be the weighted average of the medians in Table 5.2, which is MED = 99.823. Now, consider the UCL and LCL. One can take the weighted average of the R_i, $i = 1, 2, \ldots, 7$, which is 1.472. Multiply this value by the constant, A_2, conforming to the group size = 4 or 5 in Table 5.3, which is 0.729 for days 2, 3, and 6 and 0.577 for days 1, 4, 5, and 7, respectively. The UCL and LCL according to equation (5.17) for sample sizes 4 and 5 are as follows:

$$UCL_4 = 99.823 + 0.729(1.472) = 100.896, \tag{5.19}$$
$$LCL_4 = 99.823 - 0.729(1.472) = 98.750$$

and

$$UCL_5 = 99.823 + 0.577(1.472) = 100.672, \tag{5.20}$$

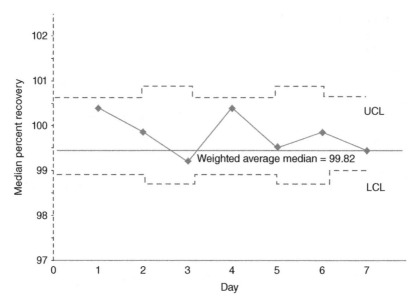

Figure 5.10   Median Control Chart for Percent Recovery (Unequal Group Sizes)

$$LCL_5 = 99.823\text{--}0.577(1.472) = 98.974,$$

respectively (Figure 5.10). Based on the medians, the process appears to be in control.

5.2.5   Mean (X-bar) and S-Charts Based on the Median Absolute Deviation (MAD)

The median absolute deviation (MAD) is often used as a measure of variability when the data is not normally distributed. We discussed the MAD with respect to outlier considerations in Chapter 4. In this section, we follow the method of Adekeye (2012), who developed the means or X-bar chart and the S-chart based on the MAD for non-normal data. Let's first set up some definitions and notations. Keep in mind that the calculations are cumbersome and can be done by hand with the appropriate tables. We demonstrate the methodology here so one can appreciate the rationale for the procedure. However, these tasks are easily done with the appropriate computer software.

Let X_{ij} represent a random sample of size n taken over m subgroups, $i = 1, 2, \ldots, n$ and $j = 1, 2, \ldots, m$. The samples are assumed to be independent and taken from a continuous identical distribution function. For example, in Table 5.1, $n = 6$ for each of the $m = 7$ subgroups. X_{ij} are the individual percent recovery values in the third column. For example, on day 1, which is the subgroup $j = 1$, we have $X_{11} = 99.8$, $X_{21} = 101, \ldots, X_{61} = 99.1$. The median for the jth subgroup is defined as

$$MD_j = \text{Median of } (X_{1j}, X_{2j}, \ldots, X_{6j}). \qquad (5.21)$$

Thus, the median of the first subgroup is $MD_1 = 100.65$, which is the average of the middle two values 100.4 and 100.9. Adekeye (2012) defines the MAD of the jth group as an adjusted median of the absolute deviations of each value in the group from the median of that group. Mathematically, this takes the form

$$MAD_j = 1.4826\,[\text{Median}|X_{ij} - MD_j|]. \tag{5.22}$$

The constant, 1.4826, is explained from a theoretical perspective by Adekeye (2012) and Abu-Shawiesh (2008) as the constant needed to make the MAD a consistent estimator for the parameter of interest under the normal distribution. In other words, in this case, the constant, 1.4862, makes the MAD an unbiased estimator of the population variance or, on average, it gives us a very close approximation to the true population variance. This makes sense, since in all that follows the MAD is used in place of the variance in the upper and lower limit calculations of our control charts.

Thus, according to (5.22), the MAD_j for each of the seven subgroups for $j = 1, 2, \ldots, 7$ are $0.667, 0.890, 0.296,\ 0.667, 0.371, 0.890$, and 0.667, respectively. The average median absolute difference is defined as

$$\overline{MAD} = \frac{1}{m} \sum_{j=1}^{m} MAD_j \tag{5.23}$$

We also require four more constants to do the calculations. They are B_3, B_4, b_n, and A_6. Like our previous constants, they depend on the sample size. B_3 and B_4 are a function of the values c_4, which we used in the earlier sections. Assuming also the usual 3 sigma units distance from the center line that we considered earlier, they are defined as

$$B_3 = 1 - (3/c_4)\sqrt{1 - c_4^2} \tag{5.24}$$

$$B_4 = 1 + (3/c_4)\sqrt{1 - c_4^2}.$$

These values and the constants of b_n and A_6 are defined in Table 5.5. Note that the values for B_3 for subgroup size less than six are not listed on the Table. These values according to (5.24) are negative. Thus, the lower limit of the S-chart formula (5.27), which we will discuss later, is set to 0. This is consistent with the calculations for the S-chart in the normal case in Figure 5.8. Also, let

$$\overline{\overline{X}} = \sum_{j=1}^{m} \overline{X}_j, \tag{5.25}$$

which is the average of the m subgroup means. We now construct the mean (X-bar) chart.

TABLE 5.5   Control Chart Constants for MAD Calculations

Control Chart Constants for MAD Calculations

Subgroup Size (n_i)	B_3	B_4	b_n	A_6
2	—	3.2665	1.196	2.5371
3	—	2.5682	1.495	2.5893
4	—	2.2661	1.363	2.0445
5	—	2.0890	1.206	1.6180
6	0.0304	1,9696	1.200	1.4697
7	0.1177	1.8823	1.140	1.2926
8	0.1851	1.8145	1.129	1.1975
9	0.2391	1.7609	1.107	1.1070
10	0.2837	1.7163	1.087	1.0312
11	0.3213	1.6787	1.078	0.9751
12	0.3535	1.6465	1.071	0.9275
13	0.3816	1.6184	1.066	0.8870
14	0.4062	1.5938	1.061	0.8507
15	0.4282	1.5718	1.056	0.8180
20	0.5102	1.4898	1.042	0.6990
25	0.5648	1,4352	1.033	0.6198

To determine the UCL, the center line which we label CL and the LCL for the means chart in a nonnormal process, we use the subscript, M, and define

$$\text{UCL}_M = \overline{\overline{X}} + A_6 \overline{\text{MAD}}$$

$$\text{CL}_M = \overline{\overline{X}} \tag{5.26}$$

$$\text{LCL}_M = \overline{\overline{X}} - A_6 \overline{\text{MAD}}.$$

To determine the UCL, the center line and the LCL for the S-chart in a nonnormal process, we use the subscript, S, and define

$$\text{UCL}_S = B_4 \, b_n \, \overline{\text{MAD}}$$

$$\text{CL}_S = b_n \, \overline{\text{MAD}} \tag{5.27}$$

$$\text{LCL}_S = B_3 \, b_n \, \overline{\text{MAD}}.$$

Adekeye (2012) notes that if the data under consideration is not normal and there is no management specification, then the control limits in equations (5.26) and (5.27) should be used to monitor the process mean and variability, respectively.

Examining equation (5.26) the values for the X-bar chart are, $\overline{\text{MAD}} = 0.6353$, $A_6 = 1.4697$ from Table 5.5, $\text{CL}_M = \overline{\overline{X}} = 100.07$ and thus, $\text{UCL}_M = 101.004$ and

$LCL_M = 99.136$. The width of the interval (LCL_M, UCL_M) is 1.87. Note that this is a slightly narrower interval than the interval derived from equations (5.1) and (5.2) when considering the data to be normal. The width of that interval (99.008, 101.135) was 2.13. As a result, we have an added percent recovery observation violating these limits from days 4 and 7 than previously noted.

Now for the S-chart based on the MAD, we have $B_3 = 0.0304$, $B_4 = 1.9696$, and $b_n = 1.2$ from Table 5.5. Thus, we have $CL_S = 0.7624$, $UCL_S = 1.502$, and $LCL_S = 0.0232$. The width of the interval (LCL_S, UCL_S) is 1.27. Note that this is a slightly narrower interval than the interval derived from equations (5.10) and (5.11) when considering the data to be normal. The width of that interval (0.025, 1.617) is 1.59. Thus, we have a more conservative estimate of the control limits. In both cases (mean chart and X-bar chart), assuming normality and nonnormality, one notes for the S-chart that all the seven values of the standard deviations from Table 5.1 are within the control limits.

So once again we note in this section that we have an alternative to constructing control charts for the mean and variance when the data is not normal. As we noted by Adekeye (2012), if the data under consideration is not normal and there is no management specification, then the control limits in equations (5.26) and (5.27) should be used to monitor the process mean and variability, respectively.

5.3   CAPABILITY ANALYSIS

Let us now return to the means chart in Figure 5.3. We discussed the process of eliminating data and reconstructing both the means chart and range chart with the reduced data set. We reconsider this strategy. Before we go about eliminating data, let's take a closer look at Figure 5.3. One interpretation of the chart is that if none of the means goes above or below either of the control limits then the process may be in control. In a laboratory situation and a manufacturing situation, you may, at times, exert more control on the process by prespecifying control limits if necessary or practical, and thus, one can consider what is termed a capability analysis, sometimes referred to as process capability or capability index.

Capability analysis is a set of calculations usually from data generated from a control chart and are used to assess whether a system is able to meet a set of specifications or requirements according to statistical criteria. However, data can be collected specifically for this purpose. Specifications or requirements are the numerical values within which the manufacturing system or laboratory is expected to operate. The specifications are set as the minimum and maximum acceptable values. Usually, there is only one set of limits, a maximum and/or minimum. Customers, engineers, supervisors, or laboratory managers usually set the specification limits. It is important to realize that specifications are not the same as control limits. Control limits come from control charts and are based on the data. Specifications are the numerical requirements of the system usually set a priori. Any method of capability analysis requires that the data is statistically stable, with no special causes of variation present – just random variation. Thus, process capability C_p, is really the ability of a process to meet what is often referred to as a product specification. C_p is defined as

$$C_p = \frac{USL - LSL}{6S},$$ (5.28)

where USL is the upper specification limit and LSL is the lower specification limit and S is the sample standard deviation. The guidelines for upper and/or lower specification vary somewhat and can depend on the application. Figure 5.11 gives an illustration of how capability looks graphically. One set of criteria for process acceptance proposed by Ekvall and Juran (1974) is that

1. $C_p < 1$ (process not adequate).

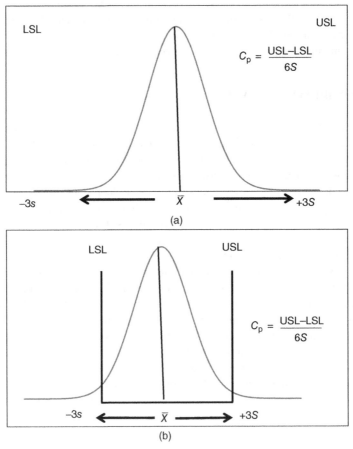

Figure 5.11 (a) Capable $C_p > 1$. (b) Capable $C_p \leq 1$

2. $1 \leq C_p \leq 1.33$ (adequate but requiring close controls as C_p approaches 1).

(5.29)

3. $C_p > 1.33$ (process is more than adequate).

In most cases, a C_p equal to 1 or greater is considered adequate. Figure 5.11a indicates that the process is adequate or better. Figure 5.11b indicates nonadequacy or worse. Note here that the lower and upper tail limits of the distribution (process) violate the LSL and USL, respectively.

Let's demonstrate the capability, C_p, by referring to our percent recovery data. Recall, we measured percent recovery of mRNA from the spiked sample. The full procedure once again is explained in Section 3.2.2. We conduct this procedure six times on each of 7 days for a total sample size of 42. Thus, we have a batch of six samples on each day, and we measure the percent recovery from each of the samples. The data is seen in Table 5.1 and in Figure 5.1. The capability calculation (5.28) examines the entire sample of 42 points. We first examine the C_p based on the control limits generated from the data. Accordingly, from the data we have LSL = LCL = 99.008 and USL = UCL = 101.135. The S or standard deviation = 0.84052. The mean (also labeled as the target, which we will explain later) is 100.0714. Once again, note here that we are using the control chart limits defined and calculated from the data, and we are not specifying any limits. Thus according to (5.28), using the control limits rather than the specified limits, we have $C_p = 0.422$, which is certainly less than one and according to (5.29), the process is inadequate. See Figure 5.12 and note how the tails of the process distribution extend beyond the LCL and the UCL. Clearly, the process violates the upper and lower control limits. Now suppose we invoke the definition of the "specified" control limits, that is, those set by the sponsor or lab manager based on expected or past performance, and we now have for example LSL = 97 and USL = 103 with $S = 0.84052$. Thus, according to (5.28), we have $C_p = 1.190$, which is certainly greater than one and according to (5.29), the process is adequate. See Figure 5.13 and note how the lower and upper tails of the process distribution are well within the LSL and the USL.

For various reasons, one may wish to determine the process capability on the lower or the upper end. For example, the quantity

$$C_{PL} = \frac{\bar{X} - LSL}{3S},$$　　　　　　　　(5.30)

where $\bar{X}$ is the sample mean, is the process capability for specifications that consist of a lower limit only (e.g., strength). Note that we are looking at the left half of the distribution so that the divisor is 3 standard deviations. We can use Figure 5.14a as an example to see once again that we have adequacy of the process ($C_{PL} > 1$). If the lower limit of the distribution curve had extended beyond the LSL, then we would have had an inadequate process or $C_{PL} \leq 1$.

Montgomery (1996) refers to C_{PL} as the "process capability ratio" for one-sided lower specifications. For this one-sided specification, the recommended minimum

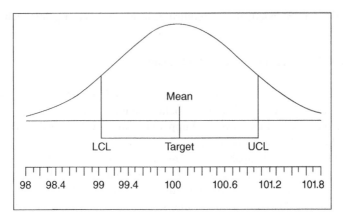

Figure 5.12   Capability Plot (CP) of Percent Recovery Data Based on the Control Limits (LCL, UCL)

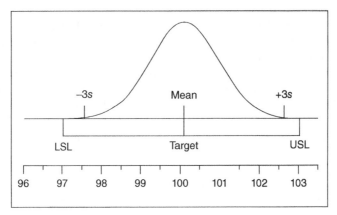

Figure 5.13   Capability Plot (CP) of Percent Recovery Data Based on the Specified Limits (LSL, USL)

values for C_{PL} are as follows:

>   1.25 for existing processes.
>
>   1.45 for new processes or for existing processes,
>
>   when the variable is critical.   (5.31)
>
>   1.60 for new processes, when the variable is critical.

A critical variable may be one that is most influential on the process or causes the most variation in a process. Such is often the case in environmental pollution monitoring. See for example, Gilbert (2007), in which control charts for total suspended particulates are presented.

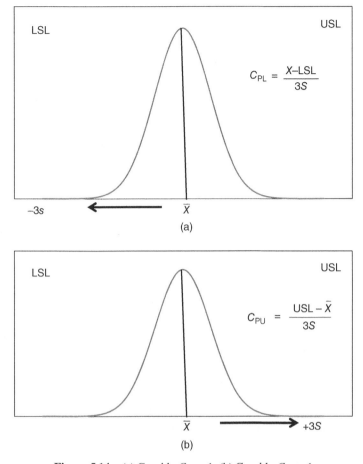

Figure 5.14 (a) Capable $C_{PL} > 1$. (b) Capable $C_{PU} > 1$

We can reexamine the left side of Figure 5.12 and compute (5.30) with the values $\overline{X} = 100.0714$, LCL = 99.009 from (5.2) and $S = 0.84052$, we have $C_{PL} = 0.421$, which is certainly less than one and not adequate. However, imposing the specified lower limit or LSL = 97 and examining the left side of Figure 5.13, then according to (5.30) we have $C_{PL} = 1.218$, which is certainly greater than one. Although the lower limits of Montgomery (1996) in (5.31) are not met, we have an indication that the process may be adequate. A specified lower limit, LSL, of 96 in Figure 5.13 would yield $C_{PL} = 1.615$, which is definitely an adequate process as per Montgomery (1996). This demonstrates how critical it is to have a stable process and also how important it is to have realistic specified limits. Also, one notes how sensitive the capability can be to these limits.

Likewise on the upper end of the process, one may have for example the quantity

$$C_{PU} = \frac{USL - \overline{X}}{3S}, \tag{5.32}$$

where $\overline{X}$ is the sample mean. C_{PU} is the process capability for specifications that consist of upper limit considerations (e.g., concentration). Note that we are looking at the right half of Figure 5.14b so that the divisor again is 3 standard deviations. We can use Figure 5.14b to see once again that we have adequacy of the process ($C_{PU} > 1$). If the upper limit of the distribution curve had extended beyond the USL, then we would have had an inadequate process or $C_{PU} \leq 1$. Montgomery (1996) refers to C_{PU} as the "process capability ratio" for one-sided upper specifications. The recommended minimum values for C_{PU} are as for the C_{PL} in (5.31). However, as noted earlier for C_P or C_{PL}, oftentimes a value greater than 1 may be deemed adequate.

To demonstrate the C_{PU} for the data generated upper limit, we can reexamine the right side of Figure 5.13 and compute (5.32) with the values $\overline{X} = 100.0714$, UCL = 101.135 from (5.1) and $S = 0.84052$, we have $C_{PU} = 0.421$, which is certainly less than one and not adequate. However, imposing the specified upper limit or USL = 103 and examining the right side of Figure 5.13, then according to (5.13) we have $C_{PU} = 1.161$, which is certainly greater than one. Although the lower limits of Montgomery in (5.31) are not met, we have an indication that the process may be adequate. A specified upper limit, USL, of 104 in Figure 5.13 would yield $C_{PU} = 1.558$, which is definitely an improved adequate process as per Montgomery (1996). This, once again, demonstrates how critical it is to have a stable process and also how important it is to have realistic specified limits.

The capability index defined as $C_{PK} = \min(C_{PL}, C_{PU})$ is sometimes used for a non-normal or a skewed process. We'll discuss a nonnormal process more formally later. In this case from our examples, using the data generated limits of LCL and UCL, we have $C_{PK} = 0.421$. This is since both C_{PL} and $C_{PU} = 0.421$. Using the lab manager or manufacturer specified limit of LSL = 97 and USL = 103, we have from our examples $C_{PK} = \min(C_{PL}, C_{PU}) = \min(1.218, 1.161) = 1.161$.

A measure that is similar to C_{PK} and takes into account variation between the process average and what we call a target value is known as C_{PM}. A target value, which we label T, is a value around which one believes the process may be centered other than the mean. The C_{PM} is defined as

$$C_{PM} = \frac{\min(T - LSL, \ USL - T)}{3[S^2 + (\overline{X} - T)^2]^{\frac{1}{2}}}. \tag{5.33}$$

Note that if, in fact, $\overline{X} = T$, then $C_{PM} = C_{PK}$. Realistically for this data, one might assume that the target is 100% recovery or $T = 100$. Then the formula (5.33) takes the form

$$C_{PM} = \frac{\min(100 - 97, \ 103 - 100)}{3[(0.8405)^2 + (100.07 - 100)^2]^{\frac{1}{2}}}. \tag{5.34}$$

Thus, in this case, obviously $\min(100 - 97, 103 - 100) = \min(3, 3) = 3$ and thus $C_{PM} = 1.185$. Accordingly, $C_{PM} > 1$ and certainly may be adequate. As one would expect and as we noted earlier, the values of USL, LSL, and in this case, the target T, are all critical. One would expect or certainly hope that the actual mean, $\overline{X}$, would be very close to the target. The closer it is to the target, the larger the value of the C_{PM}.

5.4 CAPABILITY ANALYSIS – AN ALTERNATIVE CONSIDERATION

All that we have done earlier involves an underlying assumption of the data following approximately a normal distribution. What does one do if the data is not normally distributed? One has two choices in this situation. The obvious choice is to make the data normal by a mathematical transformation; that is to say perform a mathematical function on the original set of data in the hopes that it will become normal. One can then construct control charts and perform capability analyses on the transformed data. The challenge is to be able to interpret what the new data means in the context of the problem we are considering. The alternative in the case of capability analysis is to use what we know as a nonparametric approach or statistical technique for handling data that is not normal.

We first consider a set of nonnormal data on which we want to perform a capability analysis. We revisit our percent recovery situation and we note that the data in Figure 5.15a is slightly skewed to the right. We consider transforming the data to make it normal or as "normal" as possible. The most common and perhaps the most popular set of transformations are known as the Box Cox transformations by Box and Cox (1964). Without getting into too much mathematical detail, their procedure was to identify an appropriate exponent (lambda, λ), which one uses to transform data into a normal shape. For example, transform the variable X to X^λ. One would search for values of λ between -5 and $+5$, apply them to all the data points, and determine which gives the most normal shape. Some examples of the most common exponents used are found in Table 5.6. Once again as an example, consider the data in Figure 5.15a. These are percent recoveries and one can see that they are skewed to the right. A formal statistical test of normality of the data known as the Shapiro Wilk test shows this data significantly different from a normal distribution. Figure 5.15b is the same data with a Box Cox transform $\lambda = -2.5$ and a multiplicative constant to convert it to the same units not affecting the shape. One can see that the data has a slightly larger spread, and it is less skewed but to the left because of the inverse transformation. According to Shapiro Wilk, the transformed distribution does not differ significantly from normality.

TABLE 5.6 Common Box Cox Transformations

Common Box Cox Transformations	
λ	X^λ
-1	$X^{-1} = 1/X$
-2	$X^{-2} = 1/X^2$
-0.5	$X^{-0.5} = 1/\sqrt{X}$
0	$\log(X)$
0.5	$X^{0.5} = \sqrt{X}$
2	X^2

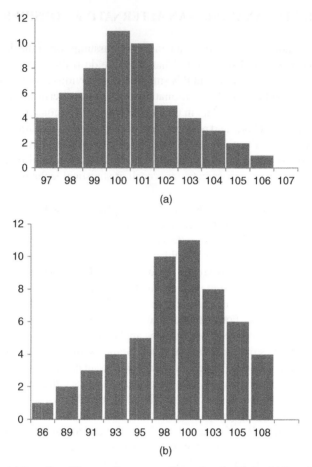

Figure 5.15   (a) Data Set of Percent Recovery – Skewed to the Right. (b) Data Set of Percent Recovery – Box Cox Transformed

Thus, if one were to perform a capability analysis on the data, then one could work with the original skewed data in Figure 5.15a or the transformed data in 5.15b. We examine the consequences of both approaches. Let's first consider the actual data in Figure 5.15a before any transformation. This is a skewed distribution and there is a nonparametric version of the C_{pk} capability statistic, which we label as C_{npk}. It has the form

$$C_{npk} = \min\left[\frac{\text{USL} - \text{median}}{p\,(0.995) - \text{median}}, \frac{\text{median} - \text{LSL}}{\text{median} - p(0.005)}\right], \qquad (5.35)$$

where $p(0.995)$ is the 99.5th percentile of the data and $p(0.005)$ is the 0.5th percentile of the data.

The median of the distribution in Figure 5.15a is 100, which is around where you would expect it to be for percent recovery experiments. Let's assume that the USL and LSL are specified as 108 and 95, respectively. In our data, $p(0.995) = 106$ and $p(0.005) = 97$. Thus, (5.13) becomes

$$C_{npk} = \min \left[\frac{108 - 100}{106 - 100}, \frac{100 - 95}{100 - 97} \right]$$
$$= \min[1.333, 1.666]. \qquad (5.36)$$

Thus, the capability $= 1.333$, which is in the realm of a capable process.

Now, according to the transformed data in Figure 5.15b, the recalculated USL and LSL translate to 114 and 82.5, respectively. The mean of the new process is 98.9 and the $S = 5.299$. Computing the $C_{PK} = \min(C_{PL}, C_{PU})$ for C_{PL} in (5.10) and C_{PU} in (5.32), we have

$$C_{PK} = \min(C_{PL}, C_{PU}) = \min(912, 840) = 0.840. \qquad (5.37)$$

Thus, according to (5.37), the capability is 0.840 and we fall short of a reasonably capable process. Clearly this does not agree with the results of the nonparametric approach in (5.36).

There are several considerations when transforming the data. First of all, you do change the data and the final form may not be meaningfully translated. In our case, the best Box Cox exponent, λ, had the value, -2.5, needed to attain the closest distribution to a normal shape. We merely multiplied the values, $PERCENT^{-2.5}$, by a constant to get values in the original units. This transformation and multiplication was also applied to the original USL $= 108$ and LSL $= 95$ to attain similar translatable values in the transformed distribution. The only change was a slight change of the mean from 100 in the original data to a value of 98.9 in the transformed data. One can probably conclude in this case that given the availability of the nonparametric approach applied to the original data, a transformation was not really necessary. The choice should be to remain with a nonparametric approach if the data is clearly not normal, unless from a previous knowledge base one knows that the data should in fact be normal.

REFERENCES

Abu-Shawiesh MOA. (2008). A simple robust control chart based on MAD. Journal of Mathematics and Statistics 4(2): 102–107.

Adekeye KS. (2012). Modified simple robust control chart based on median absolute deviation. International Journal of Statistics and Probability 1(2): 91–95. ISSN 1927–7032. E-ISSN 1927–7040.

AIAG (Automotive Industry Action Group). (2005). Manual for SPC. Code SPC-3.ISBN 97816053410889000, 2nd ed. htttp://www.buec.udel.edu/kherh/table_of_control_chart_constants.pdf.

Box GEP and Cox DR. (1964). An analysis of transformations. Journal of the Royal Statistical Society, Series B 26: 211–252.

Burr IW. (1976). Statistical Quality Control Methods, Dekker, New York.

Deming WE. (1950). Lectures on Statistical Control of Quality, Nippon Kagaku Gijutsu Remmei, Tokyo.

Deming WE. (1975). On probability as a basis for action. The American Statistician 29(4): 146–152.

Deming WE, Nippon Kagaku Gijutsu Remmei. (1952). Elementary Principles of the Statistical Control of Quality; A Series of Lectures, 2nd ed., Nippon Kagaku Gijutsu Remmei, Tokyo.

Ekvall DN and Juran JM. (1974). Manufacturing planning. Quality Control Handbook. 3rd ed., McGraw Hill, New York.

Gilbert RO. (2007). Statistical Methods for Environmental Pollution Monitoring, John Wiley & Sons. (originally published in 1987).

Montgomery DC. (1996). Introduction to Statistical Quality Control, 3rd ed., John Wiley & Sons, New York. (1st ed., 1985; 2nd ed., 1991).

Shewhart WA. (1931). Economic Control of Quality of Manufactured Product, Van Nostrand, New York. ISBN 0-87389-076-0

Shewhart WA (1939). Statistical Method from the Viewpoint of Quality Control, Dover, New York. ISBN 0-486-65232-7

6

LIMITS OF CALIBRATION

6.1 CALIBRATION: LIMIT STRATEGIES FOR LABORATORY ASSAY DATA

This chapter focuses on a comprehensive discussion of the Limits of Blank (LoB), Limits of Detection (LoD), and the Limits of Quantitation (LoQ). The comparisons of empirical and statistical approaches to these concepts are presented. Data analyses using several methodologies involving data near the detection limits are discussed. Concepts such as censored observations, extrapolation, and replacement and imputation are demonstrated. Examples of descriptive statistical data analysis include these concepts plus the trimmed mean and standard deviation and Winsorized mean and standard deviation. Hypothesis testing for this type of data are introduced as well. Robust regression on order statistics (ROS) and the Kaplan–Meier Method approach for analysis of laboratory data with nondetects are also discussed.

6.1.1 Definition – Calibration

In simple words, calibration is defined by Margaret Rouse: "In information technology and other fields, calibration is the setting or correcting of a measuring device or base level, usually by adjusting it to match or conform to a dependably known and unvarying measure." (http://whatis.techtarget.com/definition/calibration). There are challenges to quantifying an assay and its sensitivity. The (formerly) National Clinical Chemistry laboratory Standards (NCCLS) now known as the Clinical Laboratory Standards Institute (CLSI, https://www.ihs.com/products/clsi-standards.html)

Introduction to Statistical Analysis of Laboratory Data, First Edition.
Alfred A. Bartolucci, Karan P. Singh, and Sejong Bae.
© 2016 John Wiley & Sons, Inc. Published 2016 by John Wiley & Sons, Inc.

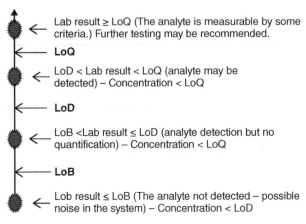

Figure 6.1   Typical Detection Results

approved guidelines to provide protocols for determination of LoD and LoQ. However, there is no consensus on these limits, and the limits are defined statistically. The LoB, LoD, and LoQ are terms used to describe the smallest concentration of a measure and that can be reliably measured by an analytical procedure. Possible instrumentation results for the LoB, LoD, and LoQ are described in Figure 6.1. We will be discussing the components of this figure as we proceed in the first few sections of this chapter.

6.2   LIMIT STRATEGIES

There are cautions and disclaimers that one must be aware of when dealing with terms such as LoB, LoD, and LoQ. Terminology, in some respects, is still in debate. Manufacturers use a wide variety of terms such as analytical sensitivity, minimum detection limits, functional sensitivity, lower limits or upper limits of detection, biological limits of detection. In this chapter, we define, for our purpose, the LoB and LoD quantitatively and their relationship to each other (CLSI).

6.2.1   Example – Estimation of LoB and LoD for Drug Assay

We give an example of blank measurements. We start with four measurements of five different blank sera, that is, patient sera without drug (Table 6.1). The measurements are listed in ascending order from 1 (lowest value) to 20 (highest value). The 95th percentile (LoB) is the 19.5th ordered measurement (the interpolated value between the values corresponding to numbers 19 and 20). In the example, we have the assay number on the left, the sorted result as the middle column, and there are four measurements per subject with the subject number in the last column. The 95th percentile is the average of the last two blank results or $(2.08 + 2.10)/2 = 2.09$ nmol/L. Therefore, by this definition, LoB $= 2.09$.

TABLE 6.1 LoB and LoD Data

Assay	Sorted Blank (nmol/L)	Subject
1	1.85	1
2	1.91	2
3	1.92	3
4	1.93	4
5	1.94	5
6	1.94	1
7	1.95	2
8	1.97	3
9	1.97	4
10	1.99	5
11	2.03	1
12	2.04	2
13	2.06	3
14	2.06	4
15	2.06	5
16	2.06	1
17	2.07	2
18	2.08	3
19	2.08	4
20	2.10	5

The LoD is defined as follows: $LoD = LoB + t \times SD_P$, where SD_P and t are, respectively, the pooled standard deviation from the five subjects and the critical value of the t-distribution discussed in Chapter 2. The SD_P is defined as the square root of the average of the variances of the five subjects. That is, $SD_P = ([SD_1^2 + SD_2^2 + SD_3^2 + SD_4^2 + SD_5^2]/5)^{1/2}$, where SD_i^2, $(i = 1, 2, \ldots, 5)$, is the variance for the ith subject. In our example, $SD_P = 0.7438$. The degrees of freedom (df) are given by $5 \times (4–1) = 15$ and the corresponding t-value is the adjusted calculation given as $1.645/(1 - 1/(4df)) = 1.645/0.9833 = 1.673$. Recall the value of 1.645 is the upper critical 95% value of the standard normal distribution. Thus,

$$LoD = 2.09 + 1.673 \times 0.7438 = 3.334. \tag{6.1}$$

From the data given in Table 6.1, LoB = 2.09 and LoD = 3.334. Note that these are in fact the results (response). We want to determine the concentration that yields the values of LoD and LoQ. The question now is, "What is the limit of quantitation, or the LoQ?" First, we need to know what the LoQ is. The LoQ is the lowest amount of the analyte in a sample that can be reliably detected and at which the total error meets laboratory requirements for accuracy or concentration at which quantitative results can be reported with a high degree of confidence (CV ≤ some value). It may

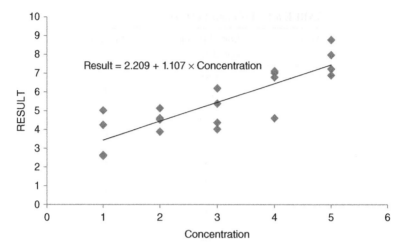

Figure 6.2   LoQ Results: Plot of Analyte Result by Concentration

in fact be the LoD. Note that the coefficient of variation (CV) is defined as the ratio of the standard deviation to the mean. It shows the extent of variability in relation to mean. Depending on the defined goal for error, the LoQ could be equal to the LoD or it could be much higher, but it cannot be lower than the LoD (LoQ ≥ LoD).

Figure 6.2 is a plot of the analyte result (not blank) versus the concentration. Note that we have four replicate samples each at five concentration levels presented in the figure. The predicted equation is given by

$$\text{Result} = 2.209 + 1.107 \text{ Concentration}. \tag{6.2}$$

Here we use the approach by Long and Winefordner (1983) and assume that the LoB is near the zero concentration level or basically the intercept of the concentration line.

6.2.2 LoQ Results

Now, we list the concentration (1–5), predicted value of the analyte, and the CV% of the five analyte values at that concentration shown in Table 6.2. Let's discuss the values of LoB and LoD. The LoB of 2.09 is below the predicted concentration value of 1 and the LoD of 3.33 is between the predicted concentration values of 1 and 2. Therefore, if we allow the lowest CV% of 5, the LoQ (which must be greater than or equal to LoD) at the CV% of 5 must be 5.41 or a concentration of 3.0. The values of LoB and LoD are indicated in Table 6.2.

This is only one approach to these quantities, and one can see the dependency of the LoQ on the smallest CV or the concentration where the CV meets a particular criteria. This obviously can be quite variable depending on definition and the experimental method. We examine other approaches to follow.

TABLE 6.2 Predicted Value of the Analyte at Each
Concentration Level and CV%

Concentration	Predicted	CV%	Limit
			LoB = 2.09
1	3.33	24.56	
			LoD = 3.33
2	4.42	25.08	
3	5.41	0.25	LoQ = 5.41
4	6.52	5.94	
5	7.74	11.6	

6.2.3 A Comparison of Empirical and Statistical Approaches to the LoD and LoQ

6.2.3.1 LoD – Statistical Approach Usually statistical and empirical methods are used to determine the LoD and LoQ. A statistical method is described to determine the LoD in Anderson (1989). Ambruster et al. (1994) compared statistical and empirical approaches to determine the LoD and LoQ for gas chromatography–mass spectrometry (GC–MS) analyses of abused drugs in the military. Among the drugs they considered were benzoylecgonine (BE), amphetamine (AMP), meth amphetamine (MAMP), codeine, and morphine. In the approach to the LoD, they consider mean value of the blank (negative blind controls) + 3 SD (standard deviation). The rationale behind using this is that the LoD should be statistically distinguishable from the blank 95–99% of the time (2–3 SDs).

6.2.3.2 LoD – Empirical Approach A series of samples containing increasingly lower concentrations of analyte is analyzed using the empirical method. One examines the lowest concentration at which the results will satisfy some predetermined acceptance criteria (Ambruster et al. 1994). As stated by the authors, they added to a series of negative urine samples certified stock solutions of drugs to produce samples with drug concentrations corresponding to serial dilutions of the cutoff calibrators. Based on these results, concentrations were selected for each drug that appeared to bracket the LoD. The means, SD, and CV were calculated from 20 replicate values. By their method, the LoD is defined as the concentration at which all routine GC–MS acceptance criteria are met 90% of the time. That is, (1) retention time within 2% of calibrator and (2) ion ratios within 20% of calibrator.

6.2.4 Example – LoD/LoQ, GC–MS Approach

We assume that there are mandated cutoff values to determine positive samples. For example, in Ambruster et al. (1994), the US Department of Defense-mandated cutoff values to determine positive samples were as follows: 100 µg/L for BE, 500 µg/L for amphetamines, and 300 µg/L for codeine and morphine, which they labeled as

TABLE 6.3   Differences between the Statistical and Empirical Results

Drug	Empirical LoD Mean	(SD) μg/L	CV%	Statistical LoD
A	2.1	(0.07)	5.1	0.51
B	7.89	(0.19)	2.4	4.7
C	132.2	(6.28)	4.7	9.63
D	78.3	(1.76)	2.2	10.23

opiates. The "target value" is measured to be equal to concentrations of the serial dilutions of the cutoff calibrators. Note that we have partial empirical (from 20 replicate values) and statistical LoD comparisons as in Ambruster et al. (1994) with each from a particular GC–MS Instrument and at least 90% of the results had to meet the acceptance criteria. In a similar experiment, and for sake of demonstration, we label like compounds as A, B, C, and D. The differences between the statistical and empirical results for our data are provided in Table 6.3.

6.2.5   LoD/LoQ, GC–MS Approach

As discussed by Ambruster et al. (1994), Long and Winefordner (1983), and others, the LoQ is recommended to be set at a higher concentration than the LoD. In order to have better probability that a value at the LoQ is not a random fluctuation of the blank reading, we consider the LoQ to be mean blank + 10 SD when one uses the statistical method (Figure 6.3).

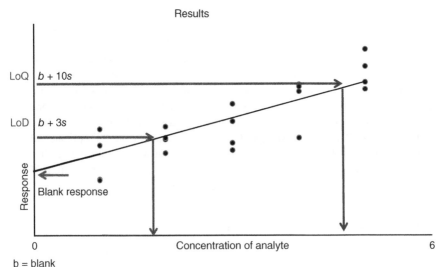

Figure 6.3   Graphical Display of Statistical LoD and LoQ

TABLE 6.4   Comparison of the Empirical and Statistical LoQ Values

Drug	Empirical LoQ (µg/L)	Statistical LoQ
A	2.70	1.71
B	9.83	12.14
C	134.2	34.67
D	85.6	21.67

The empirical method is the same procedure as above for determination of LoQ but defined as the concentration at which all acceptance criteria are met and the quantitative value is within ±20% of the target concentration (i.e., CV criteria that was earlier discussed). Table 6.4 is a partial comparison of the empirical and statistical LoQ values compared (same instruments as for LoD). Note that one points out in Tables 6.3 and 6.4 the following:

1. Any drug present in sufficient concentration to meet the LoD acceptance criteria also met the LoQ requirement of ±20% of the target value.
2. The empirical and statistical LoD and LoQ for A and B are somewhat close.
3. We note C and D exhibit a rather large difference between statistical and empirical results.

The question at this point is why do the statistical LoD and LoQ methods perform poorly?

6.2.6   Explanation of the Difficulty of the Statistical Methodology for the LoD and LoQ

Long and Winefordner (1983) and Ambruster et al. (1994) discuss the statistical method and possible reasons for its inconsistency with the GC–MS assays. It basically comes down to the imprecision with several aspects of the calibration curve including the slope or change in analyte signal as well as the intercept and concentration values. It also involves the imprecision in measuring blank values and the possible variability of exactly where the true blank measure may be. Let's examine the sources of variation.

From Table 6.1, we chose the upper 95th percentile as the blank reading. One could have just as easily chosen the mean blank value, as did Long and Winefordner (1983). First, consider the 95th percentile or 2.09 nmol/L. The variation of this reading can be determined by the method of Conover (1980) called the nonparametric 95% confidence limits for quantiles. We define the confidence limits for this quantile. Let $p = 0.95$ and our sample size is $n = 20$ and $Z_{1-\alpha/2}$ is the usual normal deviate or 1.96. The lower confidence limit for the 95th percentile is defined as follows:

$$L = p(n + 1) - Z_{1-\alpha/2}[np(1 - p)]^{1/2}. \tag{6.3}$$

And the upper limit is

$$U = p(n+1) + Z_{1-\alpha/2}[np(1-p)]^{1/2}. \qquad (6.4)$$

Solving for L and U we have

$$L = 0.95(21) - 1.96[20(0.95)(0.05)]^{1/2} = 18.04$$
$$U = 0.95(21) + 1.96[20(0.95)(0.05)]^{1/2} = 21.86. \qquad (6.5)$$

The values L and U are the order statistics from Table 6.1. The lower limit, L, is the 18th order statistic or about the blank value of 2.08. We have only 20 values. Thus, $U = 22.86$ is actually the 20th order statistic or the maximum value of 2.10. Thus, the actual blank value based on this method for future calculations could vary between 2.08 and 2.10. However, our range of value is small. For a wider range of values; this spread could be even more substantial.

Another choice of the blank value could be selected as the mean blank, which from Table 6.1 is 2.00. A 95% confidence interval on the population mean calculated as explained in Chapter 2 is 1.97–2.03. Thus, we see the contrast here in the ranges of variation depending on how the blank value is chosen. Again, our range is rather small. However, one certainly gets the idea that choice of the blank value will impact on the concentration values for the blank + 3 standard deviation values for the LoD and the blank + 10 standard deviation values for the LoQ. This is simply because the endpoints themselves, that is, the $3S$ and the $10S$ values, are random quantities. A confidence interval on the standard deviation value is defined as

$$\sqrt{\frac{(n-1)S^2}{\chi^2_{1-\alpha/2}}} < \sigma < \sqrt{\frac{(n-1)S^2}{\chi^2_{\alpha/2}}}, \qquad (6.6)$$

where $S = 0.7438$ is the sample standard deviation from above, $n = 20$ and the critical statistical values of the chi-square distribution on $n-1 = 19$ degrees of freedom at the 0.95 level are $\chi^2_{1-\alpha/2} = 32.852$ and $\chi^2_{\alpha/2} = 8.907$. Thus solving (6.6), the population standard deviation has the 95% confidence interval of 0.5657–1.0863. Thus, by simple multiplication, the $3S$ limit is between 1.697 and 3.259 and for the $10S$ limit it is between 5.657 and 10.863. Like Long and Winefordner (1983), let's examine how this affects the concentration values on the calibration curve, which we established in Figure 6.2. Solving for $3S = 1.697$, we have LoD = LoB + $3S$ = 2.09 + 1.697 = 3.787 and for $3S = 3.259$, we have LoD = LoB + $3S$ = 2.09 + 3.259 = 5.349. Using these LoD values as the result in equation (6.2) and solving for concentration, the lower and upper bounds of the concentration values are (1.53, 2.94) or effectively (2, 3). Similarly, for the lower and upper bounds of the LoQ the $10S$ limits of (5.657, 10.863) would yield concentration values of (5.11, 9.81) or effectively (5, 10). One can see here the statistical variation that may be present from this $3S$ and $10S$ approach. Actually, the $10S$ limits added to the LoB would take us beyond the upper concentration level of 5 and certainly does not make sense in this setting.

We can take this a step further. Beyond examining the effect of the concentration values based on the possible variation in the $3S$ and $10S$ limits as we just completed, we can examine the variation in the values of the concentration as Long and Winefordner (1983), taking into account the variation in the intercept and slope of the concentration curve, specifically the 95% confidence interval on the intercept and slope. Here we make use of the straight line regression principles we learned in Chapter 3, Section 3.2.1.1, and apply them to the concentration line in equation (6.2). Recall that the estimate of the intercept, β_0, is $\hat{\beta}_0 = 2.09$ and the estimate of the slope, $\hat{\beta}_1$, is 1.107. A 95% confidence interval on the intercept is defined as

$$\hat{\beta}_0 \pm t_{n-2,\ 0.975}\ S_{\hat{\beta}_0}, \tag{6.7}$$

where $t_{n-1,\ 0.975}$ is the two-sided 5% critical t-value on $n-2$ or 18 degrees of freedom since $n = 20$, which is equal to 2.101. The standard error of the estimate of β_0 is $S_{\hat{\beta}_0}$ or

$$S_{\hat{\beta}_0} = S_{X|Y}\ \sqrt{\frac{1}{n} + \frac{\overline{X}^2}{(n-1)S_X^2}}. \tag{6.8}$$

Here, $S_{X|Y}$ is the residual standard deviation or 0.7335, $\overline{X}^2$ is the mean concentration squared or 9 and S_X^2 is the concentration variance or 2.105. Thus, $S_{\hat{\beta}_0} = 0.3847$.

Therefore,

$$\hat{\beta}_0 \pm t_{n-2,\ 0.975}\ S_{\hat{\beta}_0} = 2.09 \pm 2.101\,(0.3847) \tag{6.9}$$

or a 95% confidence interval for the intercept $= (1.282, 2.898)$. Similarly, a 95% confidence interval on the slope is

$$\hat{\beta}_1 \pm t_{n-2,\ 0.975}\ S_{\hat{\beta}_1}, \tag{6.10}$$

where the standard error

$$S_{\hat{\beta}_1} = \frac{S_{X|Y}}{S_X\sqrt{n-1}}. \tag{6.11}$$

Here, $S_{X|Y}$ is defined as above and S_X is the concentration standard deviation or 1.451. Thus, $S_{\hat{\beta}_1} = 0.1160$ and

$$\hat{\beta}_1 \pm t_{n-2,\ 0.975}\ S_{\hat{\beta}_1} = 1.107 \pm 2.101\,(0.1160) \tag{6.12}$$

or a 95% confidence interval for the slope $= (0.863, 1.351)$. Thus, an upper 95% limit to the calibration curve can be expressed as

$$\text{Result} = 2.898 + 1.351 \times \text{Concentration} \tag{6.13}$$

and the lower 95% limit can be expressed as

$$Result = 1.282 + 0.863 \times Concentration. \tag{6.14}$$

Thus, using the $3S$ limits of 1.697 and 3.259 and the $10S$ limits of 5.657 and 10.863 discussed earlier, one can use the results to back calculate the concentration for the results of the values of the LoD for each of these three curves in Figure 6.4. The calculation is simply

$$Concentration = \frac{Result - \widehat{\beta}_0}{\widehat{\beta}_1}. \tag{6.15}$$

For example, for the lower limit of the LoD analyte result in equation (6.2), the concentration is

$$Concentration = \frac{3.787 - 2.09}{1.107} = 1.53, \tag{6.16}$$

as computed above and seen in Figure 6.4. Similarly, in the figure one sees that the upper limit of the LoD yields a concentration of 2.94. If we stay with these limits and back calculate for the concentration in equations (6.13) and (6.14), then we see that the concentration values for the LoD in the figure vary from 0.65 to 4.71, which is nearly the entire range of the calibration curve. Thus, by our demonstration and from Figure 6.4, one sees that the possible range of concentration values for the LoD is quite large and the statistical approach to the LoD and LoQ at times can be quite uncertain. However, it does have its place, as we'll see later in the chapter.

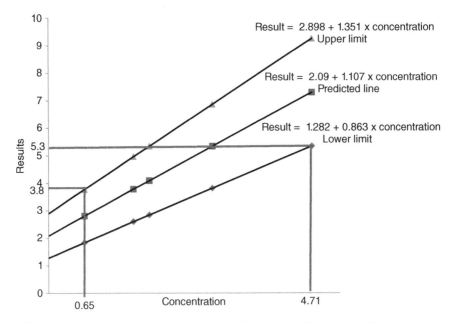

Figure 6.4   Plot of LoD versus Concentration Showing Possible Range of Concentration

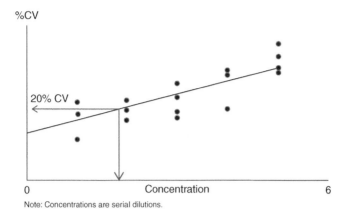

Figure 6.5   LoQ by Linear Determination

6.2.7   Another LoQ Method

The LoQ can be evaluated in some settings by determining the analyte concentration at which the pooled analyte within instrument CV was 20% on a particular instrument, A, of interest. Serial dilutions of a commercial serum-based control material or standard were prepared by dilution with normal saline or the diluent specified by the manufacturer. Each dilution was analyzed in replicates of 10 per run on several instruments, A. The LoQ was determined by plotting the %CV against the concentration and using the equation of the line to determine the concentration of analyte yielding a 20% CV. Although Figure 6.5 shows a linear relation between the %CV and concentration, this procedure has the limitation that a straight line fit may not be appropriate for the data as in this figure. One should note that concentrations are serial dilutions in Figure 6.5.

6.3   METHOD DETECTION LIMITS (EPA)

The EPA (Environmental Protection Agency 2004, 2007) defines method detection limits (MDL) as statistically determined values that define how easily measurements of a substance by a specific analytical protocol can be distinguished from measurements of a blank (background noise). The MDL is an estimate of the lower level of the calibration curve, much like the LoD we have been discussing. The MDL procedure sets the limit of detection at the 99% confidence level. We wish to calculate the MDL and LoQ in this context and verify the results. However, first we need to give some definitions.

6.3.1   Method Detection Limits

The definition of the spiked reagent blank is "a reagent blank (e.g., water) fortified with a known mass of the target analyte; usually used to determine the recovery efficiency of the method." "The best spiking level is 1–5 times the estimated (expected) MDL." (EPA 2004, 2007) The calculated MDL must be less than the spike level but greater than one-tenth of the spike level, that is,

$$\text{Calculated MDL} < \text{Spike level}. \tag{6.17}$$

However,

$$\text{Calculated MDL} > \frac{(\text{Spike Level})}{10}. \tag{6.18}$$

Let the average concentration be $\overline{X}$ and the standard deviation of results of replicate samples be S. Then we define the signal (S_{gl}) to noise (N) ratio as

$$\frac{S_{gl}}{N} = \frac{\overline{X}}{S}. \tag{6.19}$$

This describes the effect of random error on a particular measurement and estimates the expected precision of a series of measurements. Samples spiked in the appropriate range for an MDL determination typically have an S/N in the range of 2.5–10. A signal-to-noise ratio less than 2.5 indicates that the random error in a series of measurements is too high, and the determined MDL is probably high. The MDL, as we shall see, depends on the error or standard deviation, S.

The average Percent Recovery (Ave% Rec) is given by

$$\text{Ave\% Rec} = \left(\frac{\overline{X}}{\text{Spike level}} \right) \times 100\%, \tag{6.20}$$

where we use 100% of results of replicate samples.

A reasonable percent recovery is subjective and will be defined differently for different analyte situations. Since MDL calculations involve simple statistical calculations much like statistical LoD versus Empirical LoD, the recoveries may not be comparable to samples spiked somewhere well within the "quantitation" region of the calibration curve. An analyst familiar with the analytical system should be able to judge whether or not the average percent recovery falls within the expected range for relevant samples. The usual range is 80–100%.

6.3.2   Example – Atrazine by Gas Chromatography (GC)

The following is an example of the analytical work done to statistically establish the limit of detection and the limit of quantitation of atrazine in water. The analytical method used was an established EPA method. The GC was externally calibrated using five standards at 0.5, 1, 2, 5, and 10 μg/L. Ten one liter (1 L) aliquots of tap water were fortified with 0.21 μg/L of atrazine (spike). The extracts were concentrated to 1 mL, according to the established procedure. The data results are presented in Table 6.5.

TABLE 6.5   Atrazine Results

Sample Number	Results	% Recovery
1	0.19	90
2	0.22	105
3	0.20	95
4	0.20	95
5	0.19	90
6	0.24	114
7	0.23	109
8	0.18	86
9	0.17	81
Mean	0.20	96.1
Standard deviation	0.023	

We compute the MDL. By definition,

$$\text{MDL} = S \times t_{0.99,\ n-1}, \tag{6.21}$$

where S is the standard deviation.

The number of observations is equal to the number of replicates, nine, with eight degrees of freedom. The Student's t-value for nine replicates and eight degrees of freedom is 2.896 for 99% confidence. The MDL is calculated as multiplying this value by the standard deviation and rounding. Thus, from Table 6.5, MDL $= 0.023 \times 2.896 =$ 0.067. The LoQ is calculated as $10 \times S = 10 \times 0.023 = 0.23\ \mu g/L$. Now let's verify that an MDL of 0.067 is defensible:

1.	MDL < Spike or 0.067 < 0.21	OK
2.	MDL > Spike/10 or $0.067 > 0.021/10 = 0.0021$	OK
3.	$S_{gl}/N = \overline{X}/S = 0.20/0.023 = 8.69$	OK
4.	Ave % Rec = 96.1%	OK

The results 1–4 meet the requirements. Therefore, based on the above results, the MDL $= 0.067$ is reasonable.

6.3.3   LoD and LoQ Summary

The limits refer to the concentration that meets some criteria for a response to the analyte. As seen from our examples, they can be defined "slightly" differently depending on the application. Also as noted earlier, terminology in some respects is still in debate.

Manufacturers use a wide variety of terms such as analytical sensitivity, minimum detection limit, functional sensitivity, lower limits or upper limits of detection, biological limits of detection, and technical or methodological limits of detection. The statistical approach does have its limitations, but certainly is applicable in certain situations as seen in the previous section.

6.4   DATA NEAR THE DETECTION LIMITS

Now that we have discussed these limit quantities, we shall consider how they are normally reported and how to incorporate them into some statistical calculations. We define the LoD as the lowest concentration that can be determined to be statistically different from a blank. When a value is less than the LoD, however it may be defined, the analytical laboratory may report the value as:

1. Below LoD
2. Zero or LoD
3. An LT (less than) value, that is, < LoD
4. A value somewhere between zero and LoD, for example, 1/2 the LoD, or the LoD/$\sqrt{2}$ if the data is skewed.
5. The actual concentration, positive or negative, whether or not it is below the LoD. From a statistical point of view, this option, if available, may be the best, assuming the measurement is not the result of a measurement bias in the laboratory.

6.4.1   Biased Estimators

If only LT values are reported when a measurement is below the LoD, then the mean and variance of the population might be estimated by one of the following:

1. Compute the sample mean, m, and sample variance, s^2, using all measurements, including the LT values.
2. Ignore the LT values and compute m and s^2 using only the remaining "detected" values.
3. Replace the LT values by zero and compute m and s^2.
4. Replace the LT values by some value between zero and the LoD such as one half the LoD, then compute m and s^2.

6.4.2   Computing Some Statistics with the LoD in the Data

Let's introduce another term, that is, ND or "nondetects", which, like the LoD, occurs when chemical concentrations are below the detection limit of the measuring device. One can always ignore the sample mean and compute the median of the sample even with as many as half the data are NDs. This is especially relevant to skewed data. This may not be appealing to some.

The trimmed mean is an alternative to computing the median. It is also called the truncated mean because it actually eliminates points in the extremes of the distribution. Some consider it a useful estimator because it is less sensitive to outliers than is the mean but will still give a good estimate of central tendency or mean. Because it is being less sensitive to outliers, it is often referred to as a robust estimator.

The way to compute a trimmed mean is simple. We will summarize it and then demonstrate it with an example. Let p be the proportion of the sample that are LT or ND values. Compute a 100% p trimmed mean where $0 < p < 0.50$. That is, compute the arithmetic mean on the $n(1 - 2p)$ data values remaining after the largest np data values and smallest np data values are eliminated or "trimmed" away.

6.4.2.1   Example – Trimmed Mean   Suppose we have the following 10 data points arranged in ascending order ($n = 10$): 0.78, 2.3, 3.0, 3.1, 3.2, 4.1, 5.7, 6.8, 9.2, and 10.3. The mean of these 10 data points is 4.848 with a standard deviation of 2.637. Suppose the value of 0.78 is suspect. Thus, from the above summary, $p = 1/10 =$ or $2p = 0.200$. Therefore, $1 - 2p = 0.80$. This is the proportion of the sample exceeding the ND values that have been trimmed. Thus, 20% of the data is trimmed, that is, 10% from the lower end or the value, 0.78. Also, 10% of the data are trimmed from the upper end, or the value of 10.3 is trimmed. Therefore, $n(1 - 2p) = 10 \times 0.8 = 8.00 = 8$, which is the number of the data points to be analyzed after the trimming.

The remaining eight data values are 2.3, 3.0, 3.1, 3.2, 4.1, 5.7, 6.8, and 9.2, with mean = 4.675 and standard deviation of 2.689. Recall, the original mean from all 10 data points is 4.848.

6.4.2.2   Winsorized Mean   This is a technique for replacing ND values. The replacement of missing or ND values is also called imputation. Suppose we have a sample with $k \geq 1$ NDs below the LoD in a sample of n data values. The rules for Winsorizing are as follows:

1. Replace the k ND values by the next largest datum.
2. Replace the k largest values by the next smallest datum.
3. Compute the sample mean, m_w, and the standard deviation, s, of the resulting set of n data.
4. Then m_w, the Winsorized mean, is an unbiased estimator of the population mean. The Winsorized standard deviation is $S_w = s(n - 1)/(v - 1)$, where n is the total number of data values, S is the standard deviation of the new Winsorized sample, and v is the number of data not replaced during the Winsorization. The value of S_w is called unbiased. The m_w and S_w are called unbiased because, on average, they best approximate the true population mean and standard deviation of the new Winsorized sample.

6.4.2.3   Example – Winsorized Mean   Suppose we have the following data, $n = 11$:

1. ND, ND, 0.78, 2.3, 3.0, 3.1, 3.2, 4.1, 5.7, 6.8, and 9.2.
2. Ignoring the ND values, the mean is 4.242 and the standard deviation is 2.573.

3. The Winsorized data looks as follows: <u>0.78</u>, <u>0.78</u>, 0.78, 2.3, 3.0, 3.1, 3.2, 4.1, 5.7, <u>5.7</u>, and <u>5.7</u>. Here, we replace the two NDs with the value of 0.78, and then the two upper values of 6.8 and 9.2 are replaced with the value, 5.7. Thus, we have seven nonreplace values or $v = 7$

4. Thus, the new Winsorized mean is $m_w = 3.194$ and $S_w = s(n - 1)/(v - 1) = 1.949(10/6) = 3.248$.

5. Note the difference between the estimates of the means and standard deviations for the original data and Winsorized data.

   Clearly, the question is why the trimmed mean or the Winsorized mean? Like the trimmed mean, the Winsorized mean is less sensitive to outliers because it replaces them with seemingly less influential values. "Winsorization" is similar to the trimmed mean, but, instead of eliminating data, the observations are replaced or changed, allowing for some degree of influence of the replaced values. The two procedures are clearly a convenience strategy when faced with LoDs or NDs. One should keep in mind that these techniques apply and work just as well in the case that the observations appear to be above the limit of detection or you have NDs above the upper limit of instrument calibration. In this case, you begin by eliminating or replacing the upper extreme data points and then eliminate or alter that same number of points on the lower end of the distribution.

6.5 MORE ON STATISTICAL MANAGEMENT OF NONDETECTS

There is much literature on how to handle nondetects in a laboratory situation. Much is applied to EPA data (EPA 2004). What we have described earlier is basically ad hoc or substitution of the ND values in order to compute statistics such as the mean and the standard deviation. Of course, once the data is finalized, then any type of statistic may be computed. There are more sophisticated model-based approaches to handling nondetects by imputation, which we will discuss further. Of the more modern literature, Succop et al. (2004) examined data from the U.S. Department of Housing and Urban Development (HUD) sponsored lead-based Paint Hazard Control Grant Program Evaluation, which was also examined by Galke et al. (2001).

We discussed the MDL earlier. One of the findings of Galke et al. (2001) was that, using what they described as bias-corrected modeling that involves a mathematical conversion of the lead dust data, the proposed estimator from their work for the censored or ND values presumes that the median value below each laboratory's MDL is the most reliable estimate of a censored (unreported) measurement. This "most probable value," as they label this estimator, requires only the determination of the percentile at which a laboratory's MDL falls on the bias-corrected data distribution; the predicted value associated with the percentile that is half the percentile associated with the MDL is then taken as the best estimate of the censored values. For example, they note that if the MDL of a certain laboratory fell at the 10th percentile of the distribution of the data being measured, the data point associated with the 5th percentile of a normal distribution would be taken as the estimate of the censored data or the

ND value. We will discuss a methodology using percentiles of the data distribution as well. Krishnamoorthy et al. (2009) discuss an imputation approach, which involves randomly generating observations below the detection limit using the detected sample values and then analyzing the data using complete sample techniques, along with suitable adjustments to account for the imputation. This involves data simulation, which is not our goal. However, it is an interesting approach. Let's discuss some specific approaches to imputing ND or censored values that appear to be most popular.

6.5.1 Model-Based Examples of Measuring Nondetects

Robust Regression on Order Statistics (ROS)   Robust regression on order statistics (ROS) is a semiparametric method that can be used to estimate means and other statistics with censored data. ROS internally assumes that the underlying population is approximately normal. If it is not normal, then one can take the natural logarithm of the data, which may convert it into normality (often referred to as lognormal data) and one can then proceed with this method.

It is called a "semiparametric" technique, however, since the assumption is directly applied to only the censored measurements and not to the full data set. In particular, we will demonstrate that the ROS plots the detected values only on a probability plot (with a regular or log-transformed axis) and calculates a linear regression line in order to approximate the parameters of the underlying (assumed) distribution. This fitted distribution is then utilized to generate imputed estimates for each of the censored measurements, which are then combined with the known (detected) values to compute summary statistics of interest (e.g., mean and variance). The method is often called "robust" because the actual observed detected measurements are used to make estimates, rather than simply using the fitted distributional parameters from the probability plot. Let's proceed and explain the process as we go along.

Here are the steps from the EPA Unified Guidance, 2009:

Step 1: Determine the appropriate normal transformation and convert the data if necessary. Divide the data set into two groups, detects and nondetects. If the total sample size equals n, let m represent the number of detects and $(n - m)$ represent the number of nondetects. Order the m detected data from the smallest to the largest. Suppose the data points are measured in ppm and are collected in the following order:

$$5.0, \ 8.8, \ ND, \ 11.2, \ 7.7, \ 5.6, \ 6.1, \ ND, \ 5.0, \ 4.3, \ 5.0.$$

Thus, $i = 1, \dots, 11$, $n = 11$. We have $m = 9$ detects and $n - m = 11 - 9 = 2$ NDs (Table 6.6). Note that in the second column, the observations have been reordered in descending value from the top with the two NDs on the bottom, that is, 11.2, 8.8, ..., ND, ND. Notice that the nine detects are separated from the two NDs. Here, the two NDs fell below the Quantitative Limit (QL or LoD) of calibration of the measuring instrument, which was 3.0.

TABLE 6.6   ROS Method for Nine Detects and Two NDs

Observation i	Actual Concentration (ppm)	Probability $p_i = \frac{(i-0.5)}{n}$ $i = 1, \dots, n,$ $n = 11$	Standard Normal Z_i	Estimated Concentration (ppm)
11	11.2	0.954	1.68	11.2
10	8.8	0.863	1.09	8.8
9	7.7	0.772	0.74	7.7
8	6.1	0.681	0.47	6.1
7	5.6	0.591	0.23	5.6
6	5.1	0.500	0.00	5.1
5	5.0	0.409	−0.23	5.0
4	5.0	0.318	−0.47	5.0
3	4.3	0.227	−0.75	4.3
2	ND	0.136	(−1.09)	(2.6)
1	ND	0.045	(−1.70)	(0.92)

Step 2: Define the normal percentiles, p_i, for the total n sample set as follows. For a set of i values from 1 to n, $p_i = (i - 0.5)/(n)$. This notation instead of i/n for each i is known as a continuity correction adjustment (Johnson and Wichern 2007). Some texts will use the correction formula, $p_i = (i - 0.375)/(n + 0.25)$. It serves the same purpose. We will adhere to the Johnson and Wichern formula. These probabilities are calculated in the third column of Table 6.6. As an example, let's consider $i = 11$. The percentile is then

$$p_{11} = (11 - .05)/(11) = 10.5/11 = 0.954, \qquad (6.22)$$

which is the value of 0.954 in Table 6.6. Likewise, for $i = 1$ we have $(1–0.5)/11 = 0.5/11 = 0.045$, which we find in the last row of column 3 in the table. This is done for all the 11 observations.

One then converts these probabilities or percentiles into Z-values by using the inverse normal distribution. The Z-values are standardized normal variables having a mean of 0 and a variance of 1. This is not computationally straightforward, but this conversion is easily calculated using standard normal tables found in most statistical textbooks or a standard computer program.

The fourth column in Table 6.6 titled "Standard Normal Z" gives the z-values for our data. One can see that the Z_i values are separated into two groups: the larger $m = 9$ detected values in the first nine rows and the $n - m = 2$ nondetected portions in the last two rows.

Step 3: Now we consider only the $m = 9$ detected data values 4.3–11.2 in the top nine rows and use linear regression of the ordered m data values against the corresponding Z-values. In other words, regress actual concentration (ppm)

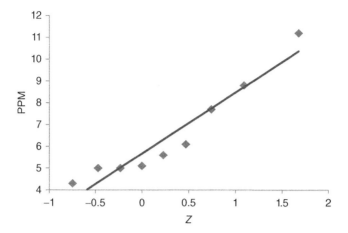

Figure 6.6   Plot of PPM versus Standard Normal Z-Values

values in the second column on the standard normal z-values in the fourth column. One then obtains the intercept and slope of the regression as the estimated mean and standard deviation using regression. The equation has the form

$$\text{Concentration (PPM)} = 5.674 + 2.801 \times Z \qquad (6.23)$$

and is seen in Figure 6.6.

Here the mean $= 5.674$ and the standard deviation $= 2.801$. However, these are based on the nine detects. Recall, our goal was to estimate or impute the two ND concentration values and then calculate the mean and standard deviation of the data. Note that the p_i values for the two NDs are 0.045 and 0.136. These result in the Z-values of -1.70 and -1.09, which using the equation (6.7) yields the imputed values

$$\text{Concentration (PPM)} = 5.674 + 2.801 \times (-1.70) = 0.92,$$

$$\text{Concentration (PPM)} = 5.674 + 2.801 \times (-1.10) = 2.6, \qquad (6.24)$$

which are in the last two rows of the estimated concentration in Table 6.6. Note that we do not use the equation to recompute the predicted values of the original $m = 9$ detects. They are known and given. That is why their values remain the same from the second to the last column in the table. The equation is used only to impute the values for the unknown nondetects or ND values, which are now 0.92 and 2.6. Now including these values with the nine detects, the mean of the 11 data points is 6.650 and the standard deviation is 2.848. We are going to revisit the ROS technique in another example, but first let's consider another way of handling the nondetects in this particular example.

6.5.2 An Alternative Regression Approach with Improvements (Refer to the Box Cox Transformation in Chapter 5)

We consider another regression approach when one wishes to impute the values below the LoD and which does not require computation of the Z-values, but does require the calculation of the percentiles as we did earlier. Here, we just consider the nine observed detected values in Table 6.6. These values are repeated in the second column of Table 6.7.

We have the same situation as earlier with two values below the LoD or two NDs. We wish to impute their values for calculations in any statistics we wish to generate, such as means and standard deviations, or perform any statistical tests. We use a predicted regression approach as earlier, but in a slightly different way. First of all, we generate the percentiles for the nine detected values only, which is the third column of Table 6.7. We proceed to get a predicted equation by regressing the actual concentration in ppm on these percentiles, p_i, $i = 1, \ldots, 9$. Thus, the Y-value is the ppm and the X-value is the actual computed percentile, p_i The predicted equation is

$$PPM = 3.128 + 6.817(p_i). \tag{6.25}$$

The plot of the data is seen in Figure 6.7. The R^2 value is 0.836. The slope is 6.817 and the intercept is 3.128. Note the similarity of the plot to Figure 6.6 when the X-values were the standard normal values. We could use equation 6.25 and proceed. However, now we become a bit suspect as one notes that perhaps one may be able to improve on the fit of the line to the points. Note in Figure 6.7 how a number of the points on the plot between the values of p_i from 0.3 to 0.8 fall below the line. We really want to have a better fit of the line to the points if at all possible. Recall in

TABLE 6.7   Alternative ROS Method for Nine Detects and Two NDs

Observation i	Actual Concentration (ppm)	Probability $p_i = \frac{(i-0.5)}{n}$ $i = 1, \ldots, n,$ $n = 9$	Predicted Concentration (ppm) for $(p_i)^{2.6}$	Residual (e_i) (Actual Predicted)	Lower 5% Limit of the Prediction Interval
9	11.2	0.944	10.857	0.342	10.105
8	8.8	0.833	9.089	−0.290	8.412
7	7.7	0.722	7.661	0.038	7.019
6	6.1	0.611	6.544	−0.445	5.912
5	5.6	0.500	5.710	−0.110	5.072
4	5.1	0.389	5.125	−0.025	4.478
3	5.0	0.278	4.755	0.244	4.101
2	5.0	0.167	4.560	0.439	3.902
1	4.3	0.056	4.494	−0.194	3.832
—	ND	—	—		(3.832)
—	ND	—	—		(3.832)

*PPM $= 4.490 + 7.396(p_i)^{2.6}$.

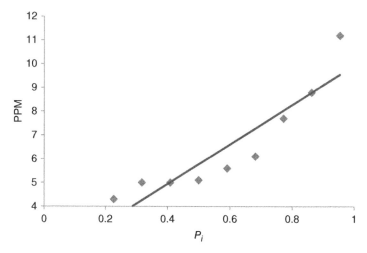

Figure 6.7 Plot of PPM versus PERCENTILES (P_i) Data

Chapter 5 that we introduced the Box Cox transformation that allows one to consider a power λ of the X-values to attempt a better fitting function to the data. We, thus, reconsider the fit in equation 6.25, and try various values of λ in the equation

$$PPM = \beta_0 + \beta_1 \ (p_i)^\lambda + \varepsilon, \tag{6.26}$$

to estimate a better fitting slope, β_1, and intercept, β_0. Usually, a power of $\lambda = 3$ or less will suffice. After attempting several power transformations, we get the following equation that gives a better fit of the line to the data:

$$PPM = 4.490 + 7.396(p_i)^{2.6}. \tag{6.27}$$

The plot of the data with the fitted line is seen in Figure 6.8. The $R^2 = 0.983$, which is much improved from 0.836 from equation (6.25). Now we can proceed with our new imputation strategy. Consider the values of p_i, $i = 1, \ldots, 9$, that is, the values in the third column of Table 6.7, which are from 0.944 to 0.056. The fourth column of the table gives the predicted PPM values according to our better fitting equation (6.27).

The last column is the lower limit of the 95% one-sided prediction interval for the values in column 4. As observation number 1 (ppm = 4.3) is the lowest value and the two NDs are below that, we wish to impute the values of the NDs using this interval approach. The rationale for the lowest limit of the interval is that the two NDs will take on the lowest 95% limit of prediction. We proceed as follows. The one-sided prediction interval on the lowest end takes the form

$$\text{Predicted (ppm)} - (t_{n-2, \ 1-\alpha}) (\text{RMSE}) \ \sqrt{1 + \left(\frac{1}{n}\right) + \frac{(X_0 - \overline{X})^2}{(n-1)\text{Var}(X)}}, \tag{6.28}$$

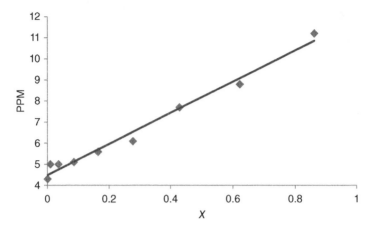

Figure 6.8 Plot of PPM versus PERCENTILES, $X=(P_i)2.6$ Data

where from Table 6.7 we have Predicted (ppm) $=4.494$. In the t-statistic we have $n-2$ degrees of freedom for $n=9$ and $\alpha=0.05$. Thus, we have the t-value of $t_{7,0.95}=1.895$. The RMSE is the root-mean-squared error or the square root of the residual variance. Recall from Chapter 3 that a residual for each of the 9 ppm observations in Table 6.7 is the difference between the actual observed observation and its value predicted by the line (equation (6.27)). For example, the residual for observation 1 in Table 6.7 is

$$e_1 = \text{Actual ppm} - \text{Predicted ppm} = 4.3 - 4.494 = -0.194. \tag{6.29}$$

The residual for each of the nine detected values is calculated and is shown in the fifth column of the table. The residual variance is computed by squaring each $e_i, i=1,\ldots,9$, summing them and dividing them by the sample size minus 2 or $n-2=9-2=7$. The RMSE is thus

$$\text{RMSE} = \sqrt{\sum_{i=1}^{n} \frac{e_i^2}{7}}. \tag{6.30}$$

The value of the RMSE here is 0.3171. Continuing to define the components of equation (6.28) we have, referring to the lowest p_i value $(i=1)$ in Table 6.7, $X_0 = (p_1)^{2.6} = (0.056)^{2.6} = 0.00056$. The value, $\overline{X}$, is the mean value of all the detected $=(p_i)^{2.6}$ for $i=1,\ldots,9$, which is 0.2764. The value Var (X) is the variance of the nine values of $(p_i)^{2.6}$, which is 0.0925. Thus, the value of the last term of equation (6.28) is

$$\sqrt{1+\left(\frac{1}{n}\right)+\frac{(X_0-\overline{X})^2}{(n-1)\text{Var}(X)}} = \sqrt{1+\left(\frac{1}{9}\right)+\frac{(0.00056-0.2764)^2}{(8)(0.0925)}} = 1.1018. \tag{6.31}$$

Thus, equation (6.28) or the lower prediction limit for the value of 4.3 in Table 6.7 is

$$4.494 - 1.895(0.3171)(1.018) = 3.832. \tag{6.32}$$

Therefore, the two nondetect points in Table 6.7 are replaced by the lowest prediction value of 3.832 and thus the values in the last column of that table.

6.5.3 Extension of the ROS Method for Multiple NDs in Various Positions

Now let's suppose the instrumentation is such that we have nondetects not only at either end such as below or above the LoD, but anywhere along the range of sampling values. Table 6.8 gives an example of 20 observations. The data was collected and then ranked in descending order. The second column gives the results of the actual values. The range is from 0.2 to 4.8. Note that in two places, we have five NDs below 3.0 and three NDs below 1.0. The strategy is to impute those eight values. We combine the imputation methodologies discussed earlier to accomplish this task.

Note that the 20th observation ($i = 20$) or actual observed data point 4.8 will have the percentile of 0.975 or $p_i = (i - 0.5)/n = (20{-}0.5)/20 = 0.975$. This yields the

TABLE 6.8 ROS Method for Multiple NDs in Various Positions

Observation i	Actual Value ppm	Probability $p_i = \frac{(i-0.5)}{n}$ $i = 1, \ldots, n, \; n = 20$	Standard Normal Z_i	Estimated Value
20	4.8	0.975	1.96	4.8
19	3.9	0.925	1.44	3.9
18	<3.0	0.875 (0.788)	(0.79)	(2.99)
17	<3.0	(0.612)	(0.29)	(2.37)
16	<3.0	(0.438)	(−0.16)	(1.84)
15	<3.0	(0.262)	(−0.64)	(1.26)
14	<3.0	(0.088)	(−0.36)	(1.61)
13	2.3	0.625	0.32	2.3
12	2.3	0.575	0.19	2.3
11	2.0	0.525	0.06	2.0
10	1.9	0.475	−0.06	1.9
9	1.6	0.425	−0.19	1.6
8	1.4	0.375	−0.32	1.4
7	1.0	0.325	−0.45	1.0
6	<1.0	0.275 (0.229)	(−0.74)	(0.929)*
5	<1.0	(0.138)	(−1.09)	(0.72)
4	<1.0	(0.046)	(−1.68)	(0.008)
3	0.6	0.125	−1.15	0.6
2	0.5	0.075	−1.44	0.5
1	0.2	0.025	−1.96	0.2

*Lower 95% prediction limit.

standard normal value of 1.96 on the first row in Table 6.8. Observation number 18 is the first nondetect and has the percentile of $(18 - 0.5)/20$ or 0.875. However, here the procedure is revised slightly, as we have five nondetects below 3.0. In this case, the value 0.875 is divided among the five values with the censoring limit below 3.0. We create new percentiles within each set on ND values to handle this situation. Some references can complicate the procedure. We simplify it here. We will proceed as follows. Since we have five NDs below 3.0, we have a new denominator of 5.0 and let $j = 1, ..., 5$. The first percentile of 0.875 for $j = 5$ is $[(5-0.5)/5] \times 0.875 = 0.788$. Similarly, for $j = 4$ we have $[(4-0.5)/5] \times 0.875 = 0.612$. We do the same for $j = 3, 2$, and 1. Note in Table 6.8, the percentiles associated with the five NDs below 3.0 range from 0.788 to 0.088. These are observations 18–14 and are indicated with parentheses. The fourth column of the table gives the standard normal values for these percentiles.

Note observation 13 or 2.3 is the next actual observed value and its percentile in the range of 20 observations is $(13-0.5)/20 = 0.625$ with a standard normal deviate of 0.32. We proceed with the observed data to observation 7 with percentile $(7-0.5)/20 = 0.325$ and normal deviate of -0.45. Note the next observation 6 is an ND. We have three NDs below 1.0. We do exactly as we did earlier. The percentile for the 6th observation out of 20 is 5.5/20 or 0.275. Since we have three NDs below 1.0, we have a new denominator of 3.0 and let $j = 1, ..., 3$. The first percentile of 0.275 for $j = 3$ is $[(3-0.5)/3] \times 0.275 = 0.299$. Similarly, for $j = 2$ we have $[(2-0.5)/3] \times 0.275 = 0.138$. We do the same for $j = 1$. Note in Table 6.8, the percentiles associated with the three NDs below 1.0 range from 0.299 to 0.046. These have normal deviates of -0.74 to -1.68 as seen in Table 6.8.

So once we have established the percentiles and their normalized values as seen in the first four columns of Table 6.8, we now wish to impute the values for the eight nondetects. To do so, we consider only the 12 detected data values 4.8, 3.9, 2.3, 2.3, 2.0, 1.9, 1.6, 1.4, 1.0, 0.6, 0.5, and 0.2 in the table and use linear regression of the ordered 12 data values against the corresponding Z-values. In other words, regress actual concentration (ppm) values in the second column on the standard normal z-values in the fourth column. One then obtains the intercept and slope of the regression as the estimated mean and standard deviation using regression. The equation has the form

$$\text{Concentration (PPM)} = 2.036 + 1.207 \times Z. \qquad (6.33)$$

The plot of the equation is seen in Figure 6.9. However, these are based on the 12 detects. Recall, our goal was to estimate or impute the eight concentration values and then calculate the mean and standard deviation of the data or other needed statistics.

Note that the p_i values for the eight NDs are seen in parentheses in the third column of Table 6.8. These result in the Z-values of the fourth column, which using the equation (6.33) yields the imputed values or estimated values in parentheses in the fifth column of Table 6.8. For example, for the Z-value of 0.79 associated with observation 18 is

$$\text{Concentration (PPM)} = 2.036 + 1.207 \times (0.79) = 2.99. \qquad (6.34)$$

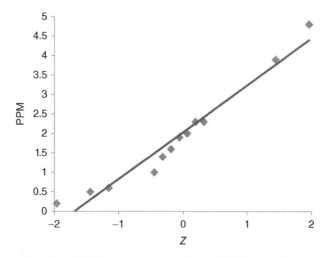

Figure 6.9 Plot of PPM versus Standard Normal Z-Values 12 Observations

Note, as before, that we do not use the equation to recompute the predicted values of the original 12 detects. They are known and given. That is why their values remain the same in the table. The equation is used only to impute the values for the unknown nondetects or ND values. Here, we introduce a caution, which is the case anytime one uses a prediction equation. Note in Table 6.8 for observation 6 that the estimated value is given as 0.929. The actual predicted value from equation 6.17 is 1.143 or $2.036 + 1.207(-0.74) = 1.143$. This is clearly above the value of 1.0. We invoked the technique introduced in the previous section. The value, 0.929, is the lower prediction limit of the equation (6.17) for the Z-value of -0.74. Thus, combining both techniques of the ROS prediction and use of the prediction limits, one has the tools to derive reasonable imputed values.

Now including these 8 values with the 12 detects, the mean of the 20 data points is 1.7114 and the standard deviation is 1.2039. The mean and standard deviation of the 12 detects are 1.875 and 1.362, respectively.

6.5.4 Cohen's Adjustment

Cohen's adjustment (1959) does not impute individual values as we did earlier. It merely uses detects and number of nondetects to compute an adjusted mean and standard deviation.

From the Unified Guidance (EPA 2009), we will outline Cohen's adjustment. Although we will not assume such here, Cohen's adjustment (Cohen 1959) has the advantage that it can be applied when as much as 50% of the observed measurements in a data set are reported as nondetects. This technique assumes that all the measurements, detects and nondetects, arise from a common population with the assumption that the lowest valued observations have been *censored* at the quantitative limit or the

QL. Using the censoring point (i.e., QL) as a location parameter that we will describe and the ordered detected values, Cohen's method attempts to reconstruct the key features of the original population, providing fairly accurate estimates of the population mean and standard deviation. These, in turn, can be used in certain statistical interval estimates and calculations where Cohen's adjusted estimates are used as replacements for the sample mean and sample standard deviation.

Cohen's adjustment assumes that the observations are sampled from an underlying normal population. The technique should only be used when the observed sample data approximately fit a normal model including transformations to normality such as lognormality. Since the presence of a large fraction of nondetects will make testing for normality difficult and there is no imputation as in the aforementioned ROS procedure, the most practical diagnostic approach is to construct a probability plot on the detected measurements as we did earlier. If the probability plot is clearly linear on the original measurement scale but not on the log scale, assume normality for purposes of computing Cohen's adjustment. If, however, the probability plot is clearly linear on the log scale, but not on the original scale, assume instead that the common underlying population is lognormal. Once it is established that the data (transformed or not) is approximately normal, then compute Cohen's adjustment to the estimated mean and standard deviation.

It is given that as the percentage of nondetects increases usually more than 50%, the accuracy of Cohen's method worsens. The Uniform Guidance (EPA 2009) does not generally recommend the use of Cohen's adjustment when more than half the data are nondetects. In such circumstances, one should consider an alternative statistical method, for instance, a nonparametric interval or perhaps the Wilcoxon rank-sum test for small samples. Here is how the procedure works. We follow the steps of the Uniform Guidance (EPA 2009):

Step 1: Divide the data set into two groups, the detects and nondetects. If the total sample size equals n, then as we did before, let m represent the number of detects and $(n-m)$ represent the number of nondetects. Denote the ith detected measurement by x_i, $i=1,\ldots,m$. Then, compute the mean and sample variance of the set of detects using the equations

$$\overline{X}_d = \frac{1}{m} \sum_{i=1}^{m} X_i \quad \text{and} \quad S_d^2 = \frac{1}{m-1} \left[\sum_{i=1}^{m} X_i^2 - m\overline{X}_d^2 \right]. \quad (6.35)$$

Let's revisit our data from our $n=11$ values of which $m=9$ are the detects in the second column of Table 6.6, that is, 11.2, 8.8, 7.7, 6.1, 5.6, 5.1, 5.0, 5.0, and 4.3. Recall that we have $n-m=2$ nondetects.

Thus, according to the expressions, 6.35, we have

$$\overline{X}_d = 6.533 \quad \text{and} \quad S_d^2 = 5.135. \quad (6.36)$$

Step 2: As is usually the case, denote the single censoring point by QL. One then computes the two derived quantities, h and Υ, necessary to derive Cohen's

TABLE 6.9   Sample of Values of λ for Cohen's Adjustment

γ				ND%			
	1	5	10	15	20	25	50
0.01	0.0102	0.0530	0.1111	0.1747	0.2443	0.3205	0.8403
0.05	0.0105	0.0547	0.1143	0.1793	0.2503	0.3279	0.8540
0.10	0.0110	0.0566	0.1180	0.1848	0.2574	0.3366	0.8703
0.20	0.0116	0.0600	0.1247	0.1946	0.2703	0.3525	0.9012
0.30	0.0122	0.0630	0.1306	0.2034	0.2819	0.3670	0.9300
0.40	0.0128	0.0657	0.1360	0.2114	0.2926	0.3803	0.9570
0.50	0.0133	0.0681	0.1409	0.2188	0.3025	0.3928	0.9826
1.00	0.0153	0.0785	0.1617	0.2502	0.3447	0.4459	1.0951
1.50	0.0170	0.0868	0.1786	0.2758	0.3793	0.4897	1.1901
2.00	0.0184	0.0940	0.1932	0.2981	0.4093	0.5279	1.2739
2.50	0.0197	0.1005	0.2062	0.3179	0.4363	0.5621	1.3498
5.00	0.0248	0.1262	0.2584	0.3977	0.5445	0.7000	1.6587

adjustment using the following equations:

$$h = 100 \times \frac{(n-m)}{n} = \text{ND\%} \quad \text{and} \quad \gamma = \frac{S_d^2}{(\overline{X}_d - \text{QL})^2}, \qquad (6.37)$$

where $h = \text{ND\%}$ is obviously the percent of nondetects. This value along with the adjustment parameter, γ, is needed for the next step. Clearly, in our case, $h = \text{ND\%} \ 100 \times (2/11) = 18.18\%$. The lower detection limit for this data is defined beforehand as $\text{QL} = 3$. Thus,

$$\gamma = \frac{S_d^2}{(\overline{X}_d - \text{QL})^2} = \frac{5.135}{(6.533 - 3.0)^2} = 0.4114. \qquad (6.38)$$

Step 3: Use the derived quantities h and γ to determine the Cohen's adjustment parameter, λ, from Table 6.9.

Clearly, the value of λ is read as a function of the values of h and γ in Table 6.9, which is a table of selected values of these parameters to fit most situations. More complete tables are found in Cohen (1959) and the Uniform Guidance (2009). For example, in Table 6.9, if $h = 15$ and $\gamma = 0.20$, then the value of $\lambda = 0.1946$, which would be inserted into the formula 6.42 of Step 4 below to compute the Cohen population adjusted mean and standard deviation. However, in our case, the values are $h = \text{ND\%} = 18.18$ or 18 and $\gamma = 0.4114$ or 0.41. One could approximate these values from the table as $h = 20$ and $\gamma = 0.40$ and derive a value of $\lambda = 0.2926$ from the table. If one is satisfied with that approach, then one would proceed as such. However, let's derive the exact value of λ for $\text{ND\%} = 18$ and $\gamma = 0.41$. This is done by means of interpolation, as follows.

We first work with the ND% and γ values that are known from the table, then put them together and derive the final λ. Note that our ND% = 18 lies between the values of 15 and 20 and $\gamma = 0.41$ lies between 0.40 and 0.50. Thus, for ND% = 15 we have

γ	λ
0.40	0.2114
0.41	L
0.50	0.2188

We want to solve for the value of λ, which we call L in this case, for the value of $\gamma = 0.41$.

This is a very simple equation, which takes the form

$$\frac{0.41 - 0.40}{0.50 - 0.40} = \frac{L - 0.2114}{0.2188 - 0.2114}. \tag{6.39}$$

Note that what we do to the left side of the equation we do exactly the same on the right side of the equation and solve for L. Here $L = 0.2121$, which makes sense as one sees it is between 0.2114 and 0.2188.

Now we do exactly the same for ND% = 20. Here we have

γ	λ
0.40	0.2926
0.41	L
0.50	0.3025

Again, we want to solve for the value of λ, which we call L in this case, for the value of $\gamma = 0.41$.

This is a very simple equation, which takes the form

$$\frac{0.41 - 0.40}{0.50 - 0.40} = \frac{L - 0.2926}{0.3025 - 0.2926}. \tag{6.40}$$

Note that what we do to the left side of the equation we do exactly the same on the right side of the equation and solve for L. Here $L = 0.2936$, which makes sense as one sees it is between 0.2926 and 0.3025. Now putting the results of (6.39) and (6.40) together, we can solve for the λ value for which ND% = 18. We, thus, have

ND%	λ
15	0.2121
18	L
20	0.2936

We want to solve for the value of λ, which we call L in this case, for the value of $\gamma = 0.41$.

This is a very simple equation, which takes the form

$$\frac{0.18 - 0.15}{0.20 - 0.15} = \frac{L - 0.2121}{0.2936 - 0.2121}. \tag{6.41}$$

Again, note that what we do to the left side of the equation we do exactly the same on the right side of the equation and solve for L. Here $L = 0.2610$, which makes sense as one sees it is between 0.2121 and 0.2936. Thus, the value of $\lambda = 0.2610$, which is certainly different than 0. 2926, we would have assumed without interpolation. Now proceed to Step 4.

Step 4: Using the values of ND% and γ derived above we use the constant, λ, from Step 3, and Table 6.9 to finalize the calculations of the adjusted estimated Cohen population mean and standard deviation of this censored data, which have the computational structures

$$\hat{\mu} = \overline{X}_d - \lambda \, (\overline{X}_d - \text{QL}) \text{ and } \hat{\sigma} = \sqrt{S_d^2 + \lambda(\overline{X}_d - \text{QL})^2}. \tag{6.42}$$

Thus,

$$\hat{\mu} = 6.533 - 0.2610 \, (6.533 - 3.0) = 5.6109. \tag{6.43}$$

$$\hat{\sigma} = \sqrt{5.135 + 0.2610 \, (6.533 - 3.0)^2} = 2.897. \tag{6.44}$$

Thus, the Cohen adjusted estimated values for the population mean and standard deviation (S) from this censored data are 5.611 and 2.897, respectively. Note that the unadjusted mean and S from just the detects were 6.533 and 2.266, respectively.

As noted, as the percentage of nondetects increases usually more than 50%, the accuracy of Cohen's method worsens. The Uniform Guidance (EPA 2009) does not generally recommend the use of Cohen's adjustment when more than half the data are nondetects. In such circumstances, one should consider an alternative statistical method, for instance, a nonparametric interval or perhaps the Wilcoxon rank-sum test for small samples.

One other requirement of Cohen's original method is that there should be just a single censoring point. Data sets with multiple censoring points, RLs (reporting limits), will usually require a more sophisticated treatment such as Kaplan–Meier, the robust ROS methods that we discussed earlier, or via maximum likelihood techniques (Cohen 1963), or perhaps a multiply censored probability plot technique (Helsel and Cohn 1988). If only 2 or 3 RLs (reporting limits) do not substantially differ and few detected intermingled data are lost, the censoring point (QL) can be set to the highest RL.

6.6 THE KAPLAN–MEIER METHOD (NONPARAMETRIC APPROACH) FOR ANALYSIS OF LABORATORY DATA WITH NONDETECTS

The Uniform Guidance (2009) does not generally recommend the use of Cohen's adjustment when more than 50% of the data are nondetects. In such circumstances, one needs to consider an alternative statistical method for analysis, for example, a nonparametric interval or perhaps the Wilcoxon rank-sum test for small samples. Also, Cohen's original method is not applicable when we have multiple censoring points. For data sets with multiple censoring points, there is a need of more sophisticated methods such as the robust ROS method that we discussed in Section 6.5.1, for example, maximum likelihood techniques (Cohen 1963), or Kaplan–Meier (a multiply censored probability plot technique) (Helsel and Cohn 1988).

Recently, nonparametric methods have received considerable attention because they don't require distributional assumptions. Nonparametric methods treating nondetects as inequalities are useful and effective techniques to handle censored data sets. The Wilcoxon rank-sum (Chapter 16.2, Unified Guidance) and Kruskal–Wallis tests (Chapter 17.2.2, Unified Guidance) are examples of rank-based nonparametric tests that can be used for data sets that contain nondetects. The reader is referred to ITRC (2013).

We discuss a nonparametric approach for calculating the cumulative probability distribution and for estimating means, sums, and variances with nondetect data (ITRC). The method is called Kaplan–Meier (Kaplan and Meier 1958). The approach is very popular for analysis of censored survival data in survival analysis. There are several types of censoring in survival analysis. In this chapter, we are interested in left censoring (nondetect measurement to the left or lower end of the distribution).

Recently, the method has been reformulated for left-censored nondetects (environmental measurements). EPA's Unified Guidance also recommends the Kaplan–Meier (KM) method for use as an intermediate step in calculating parametric prediction limits, control charts, and confidence limits for censored data sets. In this latter application, the KM estimates of the mean and variance are substituted for the sample mean and variance in the appropriate parametric formula. The KM method only determines how many data values cannot exceed any given detected level. Once the cumulative probability distribution is estimated, statistics of interest such as the mean or variance can be computed via areas under the distributional curve.

The KM method is most commonly used to calculate summary statistics such as means and variances. It can be used in conjunction with other methods to calculate upper confidence limits (UCLs) on the mean. The method can also be used to sum data that include both censored and noncensored values. This approach is often used in environmental data when calculating toxicity equivalency (TE) for dioxins and benzo (a) pyrene equivalents. The approach can also be used to improve parametric estimates of quantities such as prediction and control chart limits that require means and standard deviations and an estimate of the cumulative distribution function (CDF) properly adjusted for the presence of nondetects. This strategy is discussed in the EPA ProUCL Version 4.0 Technical Guide (2007).

Now, we lay out the formulation of the KM method for those interested. An applied numerical example follows. The process is quite cumbersome and usually done by computer. Suppose $X_1, X_2, \ldots, X_n$ are n laboratory detected or nondetected sample points. We assume that the sample points are independent. In addition, suppose $W_1, W_2, \ldots, W_k$ are k distinct values at which detects are made ($k < n$). Let m_j denote the number of detects at W_j and n_j denote the cumulative number of detects of X_i less or equal to W_j for $j = 1, 2, \ldots, k$.

Define the Kaplan–Meier (KM) probability function, P(x), (Beal 2009), as follows:

$$P(x) = \begin{cases} 1 & x \geq w_u \\ \prod_{j}^{u} (N_j - m_j)/m_j & w_1 \leq x \leq w_{u-1} \\ P(w_1) & x_1 \leq x \leq w_1 \\ 0 & 0 \leq x \leq x_1 \end{cases} \tag{6.45}$$

where u is the number of nondetects. $P(x)$ is the probability of observing a value less than or equal to a detected value.

Let $\hat{\mu}$ be the KM estimate of mean. Then, $\hat{\mu}$ is estimated by the following equation:

$$\hat{\mu} = \sum_{j=1}^{k} w_j [p(w_j) - p(w_{j-1})], \tag{6.46}$$

The KM estimate of the standard deviation (SD), $\hat{\sigma}$, is estimated by the following equation:

$$\hat{\sigma} = \sqrt{\sum_{j=1}^{k} (w_j - \hat{\mu})^2 [p(w_j) - p(w_{j-1})]}, \tag{6.47}$$

However, estimating the KM estimate of the standard error (SE) of the mean is complex. The SE is estimated by the following formulas:

$$SE = [(n - u)/(n - u - 1)] \sum_{j=1}^{k-1} C_j^2 \, m_{j+1} / [n_{j+1}(n_{j+1} - m_{j+1})], \tag{6.48}$$

where

$$C_j = \sum_{i=1}^{j} (w_{i+1} - w_i)P(w_i) \tag{6.49}$$

We can simplify equation (6.48) further by letting

$$A_j = \frac{C_j^2 \, m_{j+1}}{[n_{j+1}(n_{j+1} - m_{j+1})]}, \tag{6.50}$$

Thus,

$$SE = \left[\frac{(n-u)}{(n-u-1)} \right] \sum_{j=1}^{k-1} A_j \qquad (6.51)$$

Numerical Example on Using the Kaplan–Meier method for Analysis of Nondetect Data A laboratory scientist collected $n = 29$ sample points given as follows: 3.5, 4.0, <2.5, <2.5, 6.0, 4.5, <5.0, 6.4, 8.0, 8.0, 8.0, <9.0, 7.5, <6.5, <6.5, 7.0, <6.5, 9.5, 10.0, 12.0, 10.5, 14.0, <11.0, <14.0, <14.0, 25.0, 16.0, 75.0, 100.0.

We arrange the data points in ascending order: <2.5, <2.5, 3.5, 4.0, 4.5, <5.0, 6.0, 6.4, <6.5, <6.5, <6.5, 7.0, 7.5, 8.0, 8.0, 8.0, <9.0, 9.5, 10.0, 10.5, <11.0, 12.0, <14.0, <14.0, 14.0, 16.0, 25.0, 75.0, 100.0.

Note that

n = Number of detects and nondetects = 29

k = Number of distinct values at which detects are made = 17

u = Number of nondetects = 10

$(n - u)/(n - u - 1) = 1.0556$

Using the formulas (6.45), the calculations of the Kaplan–Meier probabilities are given in Table 6.10. Using the equation (6.49) and probabilities given in Table 6.9, we calculate $C_1, C_2, C_3, \dots, C_{16}$:

$$C_1 = (w_2 - w_1) \, P(w_1) = (4.0 - 3.5) \times 0.2077 = 0.1039$$

$$C_2 = C_1 + (w_3 - w_2) \, P(w_2) = 0.1039 + (4.5 - 4.0) \times 0.2770 = 0.2424$$

$$C_3 = C_2 + (w_4 - w_3) \, P(w_3) = 0.2424 + (6.0 - 4.5) \times 0.3462 = 0.7617$$

$$\dots$$

$$C_{16} = C_{15} + (w_{17} - w_{16}) \, P(w_{16}) = 62.6466 + (100.0 - 75.0) \times 0.9655 = 183.3341$$

Using probabilities in Table 6.10 and values of C_j, we calculate A_j:

$$A1 = (0.1039)^2 \times 1/(4(4-1)) = 0.0009$$

$$A2 = (0.2424)^2 \times 1/(5(5-1)) = 0.0029$$

$$A3 = (0.7617)^2 \times 1/(7(7-1)) = 0.0134$$

$$\dots$$

$$A16 = (183.3341)^2 \times 1/(29(29-1)) = 41.3929$$

TABLE 6.10 Calculations for Using the Kaplan–Meier Methods for Analysis of Laboratory Data with Nondetect Data

j	w_j	$w_{j-1} < x_j \leq w_j$	m_j	#ND	n_j	$P(w_j)$
17	100	100	1	0	29	$29/29 = 1.0$
16	75	75	1	0	28	$28/29 \times 1.0 = 0.9655$
15	25	25	1	0	27	$27/28 \times 0.9655 = 0.9310$
14	16	16	1	0	26	$26/27 \times 0.9310 = 0.8965$
13	14	<14.0, <14.0, 14.0	1	2	25	$25/26 \times 0.8965 = 0.8620$
12	12.5	<11.0, 12.0	1	1	22	$24/25 \times 0.8620 = 0.8276$
11	10.5	10.5	1	0	20	$21/22 \times 0.8276 = 0.7900$
10	10.0	10.0	1	0	19	$19/20 \times 0.7900 = 0.7505$
9	9.5	<9, 9.5	1	1	18	$18/19 \times 0.7505 = 0.7110$
8	8.0	8.0, 8.0, 8.0	3	0	16	$17/18 \times 0.7110 = 0.6715$
7	7.5	7.5	1	0	13	$13/16 \times 0.6715 = 0.5456$
6	7.0	<6.5, <6.5, <6.5, 7.0	1	3	12	$12/13 \times 0.5456 = 0.5036$
5	6.4	6.4	1	0	8	$11/12 \times 0.5036 = 0.4616$
4	6.0	<5.0, 6.0	1	1	7	$7/8 \times 0.4616 = 0.4039$
3	4.5	4.5	1	0	5	$6/7 \times 0.4039 = 0.3462$
2	4.0	4.0	1	0	4	$4/5 \times 0.3462 = 0.2770$
1	3.5	<2.5, <2.5, 3.5	1	2	3	$3/4 \times 0.2770 = 0.2077$

\# ND = Number of Nondetects, m_j = Number of Detects at W_j, n_j = Cumulative Number of Detects of X_i Less than or Equal to W_j for $j = 1, 2, \ldots, 17$.

TABLE 6.11 Calculated Values of C_j and A_j for Estimating the Mean, Standard Deviation, and SE for the Kaplan–Meier Analysis of Laboratory Data with Nondetects

j	w_j	#ND	m_j	n_j	$P(w_j)$	C_j	A_j
1	3.5	2	1	3	0.2077	0.1039	0.0009
2	4.0	0	1	4	0.2770	0.2424	0.0029
3	4.5	0	1	5	0.3462	0.7617	0.0138
4	6.0	1	1	7	0.4039	0.9233	0.0152
5	6.4	0	1	8	0.4616	1.2002	0.0109
6	7.0	3	1	12	0.5036	1.4520	0.0135
7	7.5	0	1	13	0.5456	1.7248	0.0429
8	8.0	0	3	16	0.6715	2.7321	0.0244
9	9.5	1	1	18	0.7110	3.0876	0.0278
10	10.0	0	1	19	0.7505	3.4628	0.0316
11	10.5	0	1	20	0.7900	4.6478	0.0468
12	12.5	1	1	22	0.8276	6.3030	0.0662
13	14.0	2	1	25	0.8621	8.0272	0.0991
14	16.0	0	1	26	0.8966	16.0466	0.3668
15	25.0	0	1	27	0.9310	62.6466	5.1913
16	75.0	0	1	28	0.9655	183.3341	41.3928
17	100	0	1	29	1.0000	—	—

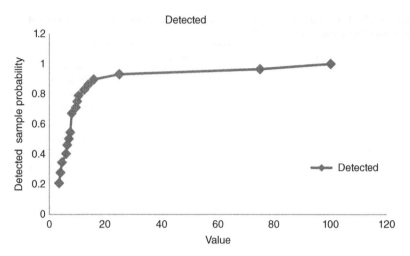

Figure 6.10   Kaplan–Meier method for Analysis of Data with Nondetects

The values of C_j and A_j are summarized in Table 6.11. Using the K–M probabilities, values of C_i given in Table 6.11, and the equations 6.46, 6.47, and 6.51 we estimate mean, standard deviation, and standard error (SE), respectively, as

$$\hat{\mu} = 17.6899,$$

$$\hat{\sigma} = 117.27,$$

$$\text{SE}(\hat{\mu}) = 1.0556 \times 47.3469 = 49.9794.$$

Now we can use the KM estimates of the mean and standard error of the mean to calculate upper confidence limits on the mean or other statistics of interest.

Additional information on how to implement the KM method and tools for calculating KM distributions and sample means can be found in the Unified Guidance, the ProUCL documentation, and Helsel (2012). Helsel (2012) includes details on calculation of the mean and standard deviation using Kaplan–Meier methods. A SAS macro for implementing the Kaplan–Meier method is available in Beal (2009). The KM curve is presented in Figure 6.10. This curve denotes the probability of detecting values less than or equal to the value on the horizontal axis.

REFERENCES

Ambruster DA, Tillman MD and Hubbs LM. (1994). Limit of detection (LOD)/limit of quantitation (LOQ): comparison of the empirical and the statistical methods exemplified with GC-MS assays of abused drugs. Clinical Chemistry 40(7): 1233–1238.

Anderson DJ. (1989). Determination of the lower limit of detection (letter). Clinical Chemistry 35: 2152–2153.

Beal D. (2009). A Macro for Calculating Summary Statistics on Left Censored Environmental Data using the Kaplan-Meier Method, Paper SDA-09, Science Applications International Corporation, Oak Ridge, TN.

Clinical and Laboratory Standards Institute (CLSI), IHS, Inc., https://www.ihs.com/products/clsi-standards.html.

Cohen AC Jr. (1959). Simplified estimators for the normal distribution when samples are single censored or truncated. Technometrics 1: 217–237.

Cohen AC Jr. (1963). Progressively censored samples in life testing. Technometrics 5(3): 327–339.

Conover WJ. (1980). Practical Nonparametric Statistics, 3rd ed., Wiley, New York.

Environmental Protection Agency (2004). Local Limits Development Guidance, Appendix Q. Methods for Handling Data Below Detection Level, Office of Water Management, EPA 833-R-04-002B.

Environmental Protection Agency (2007). ProUCL Version 4.0 Technical Guide. EPA/600/R-07/041.

Environmental Protection Agency (2009). Unified Guidance. Statistical Analysis of Groundwater Monitoring Data at RCRA Facilities. EPA 530/R-09-007.

Galke W, Clark S, Wilson J, et al. (2001). Evaluation of the HUD lead hazard control grant program: overall early findings. Environmental Research 86: 149–156.

Helsel DR. (2012). Statistics for Censored Environmental Data Using Minitab and R, 2nd ed., John Wiley and Sons, New York.

Helsel DR and Cohn TA. (1988). Estimation of descriptive statistics for multiply censored water quality data. Water Resources Research 24: 1997–2004.

Interstate Technology and Regulatory Council (ITRC), Environmental Council of States, http://www.itrcweb.org/gsmc. Documents: Groundwater Statistics and Monitoring Compliance(GSMC-1). December 2013.

Johnson RA and Wichern DW. (2007). Applied Multivariate Statistical Analysis, 6th ed., Pearson, New Jersey.

Kaplan EL and Meier P. (1958). Non-parametric estimation from incomplete observations. Journal of the American Statistical Association 53: 457–481.

Krishnamoorthy K, Mallick A and Mathew T. (2009). Model-based imputation approach for data analysis in the presence of non-detects. Annals of Occupational Hygiene 53: 249–263.

Long GL and Winefordner JD. (1983). Limit of linearity and detection for some drug abuse at the IUPAC definition. Clinical Chemistry; 55: 712A–724A.

Succop PA, Clark S, Chen M and Galke W. (2004). Imputation of data values that are less than the detection limit. Journal of Occupational and Environmental Hygiene 1: 436–441.

WhatIs.Com, Calibration, http://whatis.techtarget.com/definition/calibration.

7

CALIBRATION BIAS

This chapter discusses error, accuracy, precision, types of error, sources of error, calibration, calibration bias, uncertainty, and estimation methods of uncertainty. Some may be a review of what we touched upon in the earlier chapters. It will put us in the further context of uncertainty, which we discuss in this chapter. The chapter ends with a discussion of crude versus precise methodologies in testing the accuracy of calibration.

7.1 ERROR

Error An error is defined as the difference between a true sample point and observed sample point. For example, suppose the true sample weight is 10.5 lb and the observed sample weight by a scale is 10.0 lb. Then, the error is $10.5 - 10.0 = 0.50$ lb. Before discussing the error further in detail, we need to discuss accuracy and precision.

Accuracy Accuracy is defined as how close a measured sample point is to a known sample point. The closer the measured sample point to the known sample point, the more accurate is the measurement. For example, suppose our scale gives a sample point of 58.4 kg, but the true sample point is 75 kg. Our measured sample point is not close to the true sample point. Our sample point is not accurate. Note that accuracy

Introduction to Statistical Analysis of Laboratory Data, First Edition.
Alfred A. Bartolucci, Karan P. Singh, and Sejong Bae.
© 2016 John Wiley & Sons, Inc. Published 2016 by John Wiley & Sons, Inc.

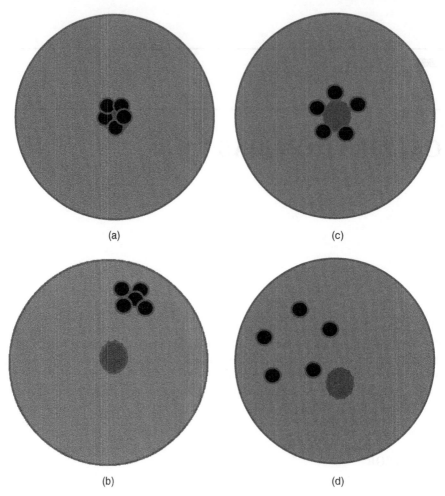

Figure 7.1 (a) Accurate and Precise: No Systematic, Little Random Error. (b) Inaccurate and Precise: Little Random Error but Significant Systematic Error. (c) Accurate and Imprecise: No Systematic, but Considerable Random Error. (d) Inaccurate and Imprecise: Both Types of Error

depends on the instrument we use, making accuracy vary from instrument to instrument. However, in general, the degree of accuracy of our instrument should be within 0.5 units of the measured sample point. For example, if our instrument measures in "1"s or units, then any sample point measured as "11" is actually between 10.5 and 11.5. In Figures 7.1a and c, the solid circles are clustered at or around the dotted circle target, illustrating accuracy by definition. Figures 7.1b and d have the black dots clustered away from the dotted circle target, illustrating inaccuracy.

Measurement of Accuracy Note that the closer the measured sample point is to the true value the more accurate the measured sample point is. Usually, we use a confidence interval to measure accuracy. For example, a confidence interval of 165.5 lb ± 0.05 for measured sample weight means that the true value lies between 165.45 and 165.55 lb as illustrated by the following diagram of interval:

Precision   Precision is defined as how close two or more sample points are to each other. Suppose we take five sample points using our instrument and get 10.5 kg each time. Then our measured sample point is very precise. Note that precision is independent of accuracy. A lab scientist may be very precise but inaccurate and vice versa. In Figures 7.1a and b, the solid circles are close to each other, illustrating precision. In Figures 7.1c and d, the solid circles are not close to each other and more spread out, illustrating imprecision. Now, let's go back to discussing error.

7.1.1   Types of Error

In general, there are three types of error often considered in laboratory data: Systematic Error, Random Error, and Erratic Error.

Systematic Error   Systematic error is caused by the equipment used in an experiment or by the observer. For example, a ruler may be giving 0.001 meters more in length because it was zeroed incorrectly. Note that systematic errors can result in high precision but have poor accuracy. Repeating the observations does not average out systematic errors. The measurement instrument is calibrated to remove systematic error. The solid circles in Figures 7.1a and c have no systematic error, but the solid circles in both Figures 7.1b and d have systematic error by their patterns.

Random Error   If an error occurs in a completely nonreproducible way from sample point to sample point, then the error is random error. Repeating experiments n times does average out and reduces random errors by a factor of $\sqrt{n}$. Statistical methods are used to deal with this type of error. For example, the random error is treated statistically by t-test and ANOVA discussed in the earlier chapters. Statistical methods relate a calculated result to the precision with which each of the experimental variables (weight, volume, etc.) is known. By looking at patterns of the black dots in Figures 7.1a and b, there is not much random error, but the black dots in Figures 7.1c and d have random error because of random patterns.

Erratic Error   An erratic error is also known as a blunder (Birch 2001). There are situations where this type of error occurs. For example, an observer may misread sample points from an instrument, or the instrument may not function properly because of poor electrical connection, displaying incorrect sample points occasionally and creating blunders. If we notice a blunder has occurred, then we must discount the readings and repeat the experiment correctly. Without noticing such blunders and without repeating the experiment correctly, erratic error can be larger than random error.

How do we know that a blunder may be the cause of measured sample points? Erratic error or a blunder occurs if a result doesn't fall within a propagated uncertainty or is larger than expected statistical uncertainty in a calculated result. This means that a result differs widely from a true or known sample point, or has low accuracy – the cause of a blunder. This situation of the cause of a blunder is also applicable to a result differing widely from the results of other experiments we have performed, or has low precision.

How do we detect erratic error or blunders? We detect erratic error or blunders by repeating all sample points at least once and to compare them to true or known sample points. Fortunately, we have rigorous statistical tests available to determine when a result or datum can be discarded because of wide discrepancy with other data in the set. We discussed this in detail in Chapter 4 concerning outlier analysis.

7.2   UNCERTAINTY

Measured and calculated sample points have uncertainty. Uncertainty may be expressed explicitly in some degree of confidence. In order to make decisions, we would like to know with some degree of confidence that the true value falls within a range of measured sample points. The range of sample points is expressed as an explicit ± uncertainty sample point. For example, a laboratory scientist gives a sample point of 2.53 ± 0.07. It is also expressed as margin of error $\pm 3\%$. That is $0.07 / 2.53 = 0.027$. The range of the sample point is from 2.46 to 2.60. That is, the scientist has some degree of confidence that the true value falls in between 2.46 and 2.60. Uncertainty may also be expressed implicitly using the appropriate significant digits. The last digit is considered uncertain in this way of expressing uncertainty. For example, the laboratory scientist reports a result of 2.53 with a minimum uncertainty of ± 0.01. That is, a range of 2.52 to 2.54. There are many sources of uncertainty. Some of the sources are named in the following section.

7.3   SOURCES OF UNCERTAINTY

In order to evaluate uncertainty, there is a need to understand the measurement process and all potential sources of the measurement uncertainty. One should study the measurement process in detail and think of possibilities of sources of uncertainty in the process. The process includes a detailed study of the measurement procedure and the measurement system. A wide variety of means, including flow diagrams and computer simulations, repeat or alternative measurements may be used in the process. Some possible sources of uncertainty include incomplete definition of the test, inadequate knowledge of the effects of errors in the environmental conditions on the measurement process, imperfect measurement of the environmental conditions, the sample may not be truly representative, personal bias in reading analog instruments, changes in the characteristics or performance of a measuring instrument since its last calibration, or errors in values of constants. Some estimation methods of uncertainty are detailed in the following section. The reader is referred to Birch (2001).

7.4   ESTIMATION METHODS OF UNCERTAINTY

How can we have meaningful comparison of equivalent results from different laboratories, within the same laboratory, or comparison of the results with reference values given in specifications or standards? This is accomplished by estimating the uncertainty of the measured sample point. We estimate the probable error, giving us an interval about the measured sample point in which we believe the true value should fall. Testing equivalence of results reduces unnecessary repetition of tests if differences are not significant and makes the process more economical.

There are several steps in estimating the uncertainty. The first step is to estimate the precision of the collected sample points. For example, let's consider the following measured sample points of concentration of a pesticide in sample of an agricultural commodity using an instrument:

1st measured sample concentration point: 1.906 ppm

2nd measured sample concentration point: 1.905 ppm

3rd measured sample concentration point: 1.907 ppm

Mean measured concentration point $= (1.906 + 1.905 + 1.907)/3$ ppm $= 1.906$ ppm

Note that the precision (reproducibility) is ± 0.001 ppm.

Therefore, all three measured sample points can be expressed as 1.906 ± 0.001 ppm. In this example, the precision is roughly 0.001 ppm or uncertainty of ± 1 in the last digit.

Now we consider different methods of estimating uncertainties in measurements. Repeating an experiment can tell us a lot about uncertainties in measurements, and we call such uncertainties random or Type A. We use statistical methods to obtain reliable estimates of Type A uncertainties through a series of observations. Other uncertainties are of Type B, for example, systematic errors or Type B of errors, which are components of uncertainties associated with errors that remain constant while a sample of measurements is taken. In the following section, we consider statistical ways of measuring Type A and Type B uncertainties.

7.4.1   Statistical Estimation Methods of Type A Uncertainty

We use statistical methods to estimate the uncertainty when we have a large number of repeated measured sample points. We use the spread in the values of many repeated sample points to estimate the uncertainty in their mean. We fit statistical models to the data. We detail a statistical method that utilizes the simple formulations we recall from Chapter 1. Suppose we have a set of M repeated independent sample points, $X_1, X_2, X_3, \dots, X_M$. We determine the mean, $\overline{X}$, by using the following equation:

$$\overline{X} = \left(\frac{1}{M}\right) \sum_{i=1}^{M} X_i, \tag{7.1}$$

where X_i are the individual results.

The standard deviation, S, is a measure of how close the measured sample points are to the mean, and this measure is calculated by taking the square root of S^2, variance. First we calculate the variance using the following equation:

$$S^2 = \sum_{i=1}^{M} \frac{(X_i - \overline{X})^2}{(M-1)}. \tag{7.2}$$

$$S = \sqrt{S^2}. \tag{7.3}$$

Now, we calculate the standard error (SE) of the uncertainty of the mean as

$$SE = \frac{S}{\sqrt{M}}. \tag{7.4}$$

In equation (7.4), the SE depends on M. This means the more the number of measurements, the less the standard error of the uncertainty of the mean. The reader is referred to Chapter 1 for more details. The S and SE are the contributions to uncertainty.

Finally, we calculate a $(1-\alpha)$ 100% confidence interval for the true mean of the uncertainty, where α is the probability of the statistical type I error. Note that the type I error is rejecting the mean of the uncertainty, when in fact the mean is the true value in the population. In general, we assume α of 0.05. Then we can say that with $(1-0.05)$ 100% = 95% confidence that the true mean of the uncertainty lies within the range of the confidence interval (CI) given that there are no systematic errors. We calculate the confidence interval using the following equation:

$$CI = \overline{X} \pm SE \times t, \tag{7.5}$$

where t is the value of the t-statistic, sometimes called the coverage factor or critical value, for the number of the uncertainty mean, the number of degree of freedom $(M-1)$, and the confidence interval as discussed in Chapter 2.

Example: Let's consider four measured sample points of concentration of a pesticide in samples of an agricultural commodity by using an instrument:

1st measured sample concentration point: 1.510 ppm
2nd measured sample concentration point: 1.515 ppm
3rd measured sample concentration point: 1.512 ppm
4th measured sample concentration point: 1.518 ppm

To review, we calculate the mean value, variance, standard deviation, and standard error as follows:

$$\overline{X}(\text{mean value}) = \frac{(1.510 + 1.515 + 1.512 + 1.518)}{4} = 1.514 \,\text{ppm}.$$

$$S^2(\text{variance}) = [(1.510 - 1.514)^2 + (1.515 - 1.514)^2 + (1.512 - 1.514)^2$$
$$+ (1.518 - 1.514)^2]/(4 - 1)$$
$$= (0.000037)/3$$
$$= 1.233 \times 10^{-05} \,\text{squared ppm}.$$

$$S(\text{standard deviation}) = \sqrt{S^2}$$
$$= 3.5114 \times 10^{-3} \,\text{ppm}.$$

$$SE \ (\text{standard error}) = S/\sqrt{M}$$
$$= 1.7557 \times 10^{-3} \,\text{ppm}.$$

Now we calculate a 95% confidence interval for the true mean using equation (7.5). Note that the t-value or coverage factor from t-tables given in most statistics books is 3.18 with corresponding $4 - 1 = 3$ degrees of freedom for 95% confidence. Thus,

$$CI = \overline{X} \pm SE \times t$$
$$= 1.514 \pm 1.7557 \times 10^{-3} \times 3.18$$
$$= 1.514 \pm 5.5831 \times 10^{-3}$$
$$= (1.514 - 5.5831 \times 10^{-3}, 1.514 + 5.5831 \times 10^{-3})$$
$$= (1.5084, 1.5196) \,\text{ppm}.$$

We are 95% confident that the true value of the concentration of the pesticide lies between 1.5084 ppm and 1.5196 ppm.

7.4.2   Estimation Methods of Type B Uncertainty

The Type B uncertainty (u) is a set of components of uncertainties associated with errors that remain constant while a sample of measurements is taken. This uncertainty depends on scientific judgment using all the available information. It is not necessarily from a series of measurements as we've seen earlier. The information may include manufacturer's specifications, data provided in calibration reports, and knowledge of the measurement process (Taylor and Kuyatt 1994). Also, the uncertainty that changes as a function of one or more variables may be a combination of

Type A and Type B (Reda, NREL Technical Report, NREL/TP-3B10-52194, July 2011). Thus, we may use statistical, nonstatistical, or both methods for estimating the Type B uncertainty. In order to use a t-value for estimating an interval with some degree of confidence, we assume the normal distribution.

Considering some partial statistical approaches to measuring uncertainty, the Type B approach often involves a measure of uncertainty that defines an interval within which one believes the particular measure of interest will occur with some degree of confidence. The measure of uncertainty under this condition is called an expanded uncertainty. It is not a formal confidence interval approach discussed earlier, but rather involves a statement of belief about the range of the value. For example, let y be the result of a quantity we wish to measure, Y. For example, Y may be temperature and the data result we measure is y. The value Y is often called the measurand. It could be temperature, mass, humidity, and so on. Suppose we have an uncertainty value, $u(y)$, and given coverage factor, k, then the expanded uncertainty often denoted by U is defined as $k\,u(y)$. That is to say, one believes that Y is greater than or equal to $y - U$ and is less than $y + U$, which is commonly written as

$$Y = y \pm U. \tag{7.6}$$

As an example, suppose one assumes an underlying normal distribution of a process with a standard of temperature, which is $T_s = 80\,°C$ with a standard uncertainty of $u_T = 5\,°C$ and a coverage factor $k = 2$, then

$$U = k\ u_T$$
$$= 2 \times 5 = 10°C. \tag{7.7}$$

We then express our result as

$$T_s \pm U = T_s \pm k\ u_T$$
$$= (80 \pm 10)°C \tag{7.8}$$

Since it is assumed that the possible estimated values of the standard are approximately normal with approximate standard deviation of u_T, then the unknown value of the standard is believed to lie in the interval defined by U with a level of confidence of approximately 95%. The critical value of a normal distribution for 95% confidence is, in fact, 1.96.

In the absence of a value for the coverage factor, a factor of 2 may be used if the expanded uncertainty has a 95% confidence level. In many applications, u is the standard uncertainty of the calibration value given on the calibration certificate and U would be the expanded uncertainty at a 95% confidence limit on the certificate.

Moving away from the normal assumption, let's suppose that we stay with our aforementioned temperature example and we know the range of temperatures in a process but assume that there is no specific knowledge about possible values within that range. Let's label the lower limit of the range value as a_0 and the upper range

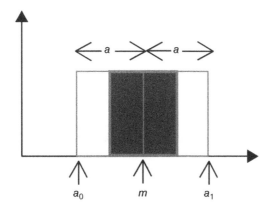

Figure 7.2   Rectangular Distribution

value as a_1. The additional assumption (EPA 1997) is that any temperature within that range occurs with equal or uniform probability. Thus, the data is in what is described as a uniform or rectangular distribution (Figure 7.2). Therefore, the mean value, m, of any random quantity, X, in the distribution is defined as the midpoint of the interval or

$$m = \frac{(a_0 + a_1)}{2},$$ (7.9)

with variance (squared uncertainty) defined as

$$u^2(x) = \frac{(a_1 - a_0)^2}{12}.$$ (7.10)

If the difference between the bounds, a_1 and a_0 is $2a$, then $u^2(x)$ becomes

$$u^2(x) = \frac{a^2}{3}.$$ (7.11)

These results are directly from the definition of the mean and variance of a uniform or rectangular distribution.

As an example, if little is known about the specific temperature inputs of a process, but that the range is from $a_0 = 95\,°C$ to $a_1 = 105\,°C$ we, thus, have $2a = (105 - 95)$ or $a = 5\,°C$. Thus, the standard uncertainty takes the value

$$u(x) = a/\sqrt{3}$$
$$= 5/\sqrt{3} = 2.9\,°C.$$ (7.12)

The rectangular distribution that we utilize above assumes equidistance of the end-points of the interval to the midpoint, or what we usually describe as a symmetric distribution. The input quantities may not be symmetric about the midpoint, X. That is to say, if we define the lower bound as $a_0 = x - L$ and the upper bound as $a_1 = x + U$, where $L \neq U$, clearly x is not in the center of the interval, a_0 to a_1 and thus the distribution cannot be uniform. However, there may be no other reasonable choice for a distribution. In that case, the approximation to the squared uncertainty is given as

$$u^2(x) = \frac{(U+L)^2}{12} = \frac{(a_1 - x + x - a_0)^2}{12} = \frac{(a_1 - a_0)^2}{12}. \tag{7.13}$$

This is the variance of a rectangular distribution with width equal to $U + L$. Given our aforementioned temperature example, the standard uncertainty is defined as

$$u(x) = \frac{(a_1 - a_0)}{\sqrt{12}} = \frac{(105 - 95)}{\sqrt{12}} = \frac{10}{\sqrt{12}} = 2.9\,°C. \tag{7.14}$$

Clearly, nonsymmetry of a_0 and a_1 in Figure 7.2 will yield a different result.

Another option is not to assume a rectangular distribution to start with. There may be no rationale for assuming a uniform distribution of data points between two bounds. It may be more reasonable to assume that the values near the bounds are less frequent than those near the midpoint. Here, one assumes a triangular distribution as seen in Figure 7.3, which has sloping sides. This is sometimes referred to as a trapezoidal distribution. Here, like the aforementioned rectangular example, one assumes the half width mark at $a = (a_1 - a_0)/2$. Thus, the standard uncertainty is defined as

$$u(x) = \frac{a}{\sqrt{6}}. \tag{7.15}$$

Again, using our temperature data as above, clearly $u(x) = a/\sqrt{6} = 5/2.45 = 2.04\,°C$.

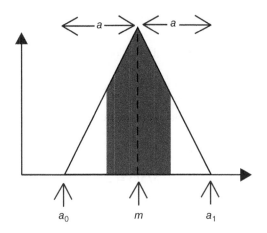

Figure 7.3   Triangular Distribution

The estimate of standard uncertainty using the triangular distribution is smaller than the estimate obtained using the rectangular distribution. Theoretically, this is what was expected.

7.4.3 Estimation Methods of Combined and Expanded Uncertainties (Normal Data)

The first step is to identify, estimate, and express the uncertainty components as standard deviations. If a measure contains all random or systematic uncertainty contributions, then we combine these estimates all together for estimation of combined uncertainty after converting the measure of dispersion into the standard deviation. However, if we have repeated measurements, then the dispersion is already expressed in the form of a standard deviation (EPA 1997). We calculate the standard deviation of a simple uncertainty estimated from specification sheets with a level of confidence ($\pm\alpha$ some probability level) by dividing the expanded uncertainty, $k\,u_T$, by the appropriate divisor corresponding to the given confidence interval. For example, we use divisors of $k = 1, 2, 3$ for confidence intervals 68%, 95%, and 99%, respectively, and by assuming a normal probability distribution. We use divisors $\sqrt{3}$ and $\sqrt{6}$ for rectangular and triangular probability distributions, respectively, as we saw in the previous section.

The next step is to calculate the combined standard deviations $u(y)$, which are expressed in terms of sensitivity coefficients, $C_1, C_2, C_3, \ldots, C_n$, and the uncertainty of the independent sample points, $X_1, X_2, X_3, \ldots, X_n$, as

$$u(y) = \sqrt{\sum_{i=1}^{n} C_i^2 u(X_i)^2}, \tag{7.16}$$

where the sensitivity coefficients are the partial derivatives of $u(X_i)$ with respect to the uncertainty components X_i. The law of propagation of uncertainty is also expressed by (7.16). The reader is referred to Barr and Zehna (1983). However, although the notation in (7.16) looks foreboding, we are able to construct such an example here. Table 7.1 contains a small sample of air pollution data measured early afternoon in a large American city. The ozone (Y) is measured in Dobson Units (DU), the solar radiation (X_1) in watts/meter^2, and the NO_2 (X_2) in ppb. The investigators were examining the linear relationship of the ozone to the solar radiation and NO_2. The standardized uncertainties are much like the regression coefficients discussed in Chapter 3.

The fitted regression equation is

$$Y = 0.081\ X_1 + 0.460\ X_2. \tag{7.17}$$

Here, $C_1 = 0.081$ and $C_2 = 0.460$. The values corresponding to the contribution to the uncertainty of X_1 and X_2 are their standard deviations or $u(X_1) = 13.141$ and

TABLE 7.1 Normally Distributed Air Pollution Data

| | Environmental Data (Normal) | | |
Observation	Ozone (Y)	Solar Radiation (X_1)	NO$_2$ (X_2)
1	7	75	8
2	12	81	14
3	4	45	6
4	21	82	11
5	7	72	8
6	12	77	12
7	11	65	7
8	9	61	8
9	3	47	6
10	8	72	11

$u(X_2) = 2.726$. Thus, solving equation (7.16), we have

$$u(y) = \sqrt{\sum_{i=1}^{n} C_i^2 u(X_i)^2} = \sqrt{0.081^2(13.141)^2 + 0.460^2(2.726)^2}$$

$$= 1.645. \tag{7.18}$$

Thus, based on the contributions of the solar radiation and the NO$_2$, the combined uncertainty of the ozone is 1.645. This is a very simple example involving only two components ($n = 2$), that is, X_1 and X_2. The generalization to more components is straightforward as seen from equation (7.16).

Sometimes when considering combined uncertainties, the X components may not be independent. In fact, they may be correlated, and the correlation may be significant. One would have to include this information in equation (7.16), and it can become very cumbersome mathematically. In our aforementioned example, we generated equation (7.17) under the assumption that solar radiation and NO$_2$ were independent of each other. They are, in fact, correlated, and positively so. That is to say, in this small data set of Table 7.1, as the solar radiation increases the NO$_2$ increases. that is, they change in the same direction. This is seen in Figure 7.4. If they changed concurrently in the opposite direction, that is, as one increases the other decreases, then the correlation would be negative.

The value of the correlation between these two variables is 0.814. The correlation in general takes values between -1 and $+1$; the values in between measure the strength of the correlation. Values tending to -1 indicate a strong negative correlation and values tending to $+1$ indicate a strong positive correlation. The covariance is another measure of how the variables change together and it behaves like the correlation, but without a restricted upper or lower bound of 1. The sample covariance

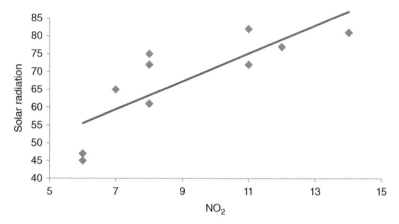

Figure 7.4 Solar Radiation versus NO_2

between two variables, X_1 and X_2, is simply defined as

$$\text{Cov } (X_1, X_2) = \sum_{i=1}^{n} (X_{i1} - \overline{X}_1)(X_{i2} - \overline{X}_2), \tag{7.19}$$

where from our notation above and in Chapter 1, $n = 2$, $\overline{X}_1$ and $\overline{X}_2$ are the sample means of the solar radiation and NO_2, respectively, or from Table 7.1 $\overline{X}_1 = 67.70$ and $\overline{X}_2 = 9.10$. Doing the calculation in (7.19) results in the covariance of 29.144. We will incorporate this information into equation (7.16) to account for the dependence between solar radiation and NO_2 and determine how the combined uncertainty value changes. However, before we do that, as an added note, let's examine the relationship between the covariance and the correlation. The correlation of X_1 and X_2 or Corr (X_1, X_2) is simply defined as the covariance of X_1 and X_2 divided by the product of their standard deviations, which are 13.141 and 2.726. Thus,

$$\text{Corr } (X_1, X_2) = \frac{29.144}{(13.141)(2.726)} = 0.814. \tag{7.20}$$

So it is obvious that there is a strong positive correlation or covariance between solar radiation and NO_2. Therefore, we incorporate the covariance into equation (7.16) as follows:

$$u(y) = \sqrt{\sum_{i=1}^{n} C_i^2 u(X_i)^2 + C_1 C_2 u(X_1, X_2)}, \tag{7.21}$$

where $u(X_1, X_2)$ is just the $\text{Cov}(X_1, X_2)$ and C_1 and C_2 are the sensitivity coefficients we already calculated in (7.17) and $n = 2$. Thus, computing (7.21), we have a new combined measure of uncertainty incorporating the covariance between X_1 and X_2, which is

$$u(y) = \sqrt{0.081^2(13.141)^2 + 0.460^2(2.726)^2 + (0.081)(0.460)(29.144)}$$

$$= 1.947. \tag{7.22}$$

Therefore, based on the contributions of solar radiation and NO_2 and the dependence between them, the combined uncertainty of the ozone is 1.947. Note that without the covariance information, the uncertainty measure was 1.645. Thus, in this case, it increased from the assumption of independence between solar radiation and NO_2. Also, if we had other factors to include calculating the combined uncertainty such as carbon monoxide, nitrous oxide, and wind, equation (7.21) would be much more complicated with n = number of components and the expression to the right side under the square root in (7.21) would be the sum of all pairwise covariances of the components (Eurachem/CITAC Guide 2000). For our purposes, an example with just two components suffices to demonstrate the calculations involved. For many components, a computer program would be needed.

7.4.4 Estimation Methods of Combined and Expanded Uncertainties (Nonnormal Data)

The estimation methods of Section 7.4.3 can be repeated if the data sample points are not normal and perhaps take on a skewed characteristic as is seen in Table 7.2. This is the same type of environmental data relating solar radiation (X_1) and NO_2 (X_2) to the ozone (Y), but note in the table that the values for each variable are skewed to the left. Here we can follow the procedure of Chapter 1 and take the log transformation of each of the variables, that is, $Z_1 = \ln(X_1)$, $Z_2 = \ln(X_2)$, and $W = \ln(Y)$. This will yield an approximate log normal distribution for each of the variables. Using these quantities, we can, thus, compute the geometric standard deviation (GSD) of each of the variables. Recall from Chapter 1 that the $\ln$ (GSD) of a set of variables = arithmetic SD of the log transformed variables.

For the sake of demonstration, we will deal with log units. The regression equation is now

$$W = -0.510Z_1 + 2.00Z_2. \tag{7.23}$$

The standardized uncertainties are much like the regression coefficients discussed in Chapter 3. Here $C_1 = -0.510$ and $C_2 = 2.00$. The values corresponding to the contribution to the uncertainty of Z_1 and Z_2 are their GSDs or $u(Z_1) = 0.088$ and $u(Z_2) = 0.313$. These are derived as the standard deviations of the log transformed values in Table 7.2. Thus, solving equation (7.16), we have

$$u(w) = \sqrt{\sum_{i=1}^{n} C_i^2 u(Z_i)^2} = \sqrt{(-0.510)^2(0.088)^2 + 2.00^2(0.313)^2}$$

$$= 0.627. \tag{7.24}$$

TABLE 7.2   Nonnormally Distributed Air Pollution Data

	Environmental Data (Nonnormal)		
Observation	Ozone (Y)	Solar Radiation (X_1)	NO_2 (X_2)
1	21	81	15
2	22	81	11
3	12	80	11
4	22	82	15
5	7	76	8
6	9	76	8
7	8	75	7
8	4	74	7
9	7	65	7
10	3	64	7

Thus, based on the contributions of the log transformed solar radiation and NO_2, the combined uncertainty of the Ozone is 0.627. Examining the covariance structure between Z_1 and Z_2, we have

$$\text{Cov } (Z_1, Z_2) = 0.021, \tag{7.25}$$

which yields

$$\text{Corr } (Z_1, Z_2) = \frac{0.021}{(0.088)(0.313)} = 0.762. \tag{7.26}$$

So it is obvious that there is a strong positive correlation between the transformed solar radiation and NO_2. Thus, we incorporate the covariance into equation (7.16) as follows:

$$u(w) = \sqrt{\sum_{i=1}^{n} C_i^2 u(Z_i)^2 + C_1 C_2 u(Z_1, Z_2)}, \tag{7.27}$$

Therefore, computing equation (7.27), we have a new combined measure of uncertainty incorporating the covariance between X_1 and X_2, which is

$$u(y) = \sqrt{0.510^2 (0.088)^2 + 2.00^2 (0.313)^2 + (-0.510)(2.00)(0.021)}$$

$$= 0.608. \tag{7.28}$$

Thus, based on the contributions of the log transformed solar radiation and the NO_2 and the dependence between them, the combined uncertainty of the log Ozone is 0.608. Note that without the covariance information the uncertainty measure was 0.627. Thus, in this case, as in the previous section, it did change somewhat from the assumption of independence between solar radiation and NO_2.

7.4.5   Another Method of Estimating Uncertainties for Nonnormal Data (Nonparametric)

Here we discuss a common statistical method called the Bootstrap (Effron and Tibshirani, 1993) for handling estimation from distributions that are not normally distributed or have no other underlying known distributional shape. This is certainly common in the laboratory setting. The method is computer intensive and was discussed in detail by Efron and Tibshirani (1993). The basic idea is to resample the data many times from the original sample and establish reasonable estimates from these many samples. Thus, in practical applications, the bootstrap means using some form of resampling with replacement from the actual collected data, to generate many bootstrap samples.

Usually, if the data (sample) consists of n independent observations or data points, then a resampled size of n with replacement is taken from the original n observations, to get one bootstrap sample. This is done repeatedly for as many bootstrap samples as one desires. However, the nature of the correct bootstrap data resampling can be more complex for more complex data structures. We keep our example simple.

If we choose to generate $K = 100$ bootstrap samples, then the set of generated bootstrap samples (say 100) is considered a surrogate or proxy for a set of 100 independent real samples. In reality, we have only one actual data sample. Estimated values of parameters from replicate bootstrap samples are derived from the bootstrap samples by analyzing each bootstrap sample exactly as we first analyzed the actual data sample. From the set of results of sample size K, we estimate our uncertainties in the present case or any other parameter of interest. We demonstrate with an example using a small sample of 11 NO_2 values. Keep in mind that the Bootstrap is best suited for larger samples of perhaps greater than 30 or 40. Here is how it works.

Consider a small sample of NO_2 ($n = 11$) values as follows:

$$9, \quad 14, \quad 10, \quad 11, \quad 11, \quad 10, \quad 9, \quad 9, \quad 9, \quad 13, \quad 10 \qquad (7.29)$$

This data is skewed to the right with a Pearson skewness value of SMC = 1.225 or moderate skewness (Chapter 2). The data set is clearly not normally distributed. The sample median of these data = 10.00 with an interquartile range (IQR) of 9–11, which in the nonparametric setting is most representative of the center of the distribution. Suppose we wanted to express an uncertainty measure in terms of the median, the lower (25th) quartile, and the upper (75th) quartile. This is certainly not the usual approach like examining an uncertainty in terms of the standard deviation of the sample, but the power of the bootstrap allows us to create uncertainty with reference to any parameter, and we, thus, demonstrate the bootstrap process. First, we draw a random subsample of size 11 with replacement from the data given in (7.29), that is, once we (actually the computer) select a value from (7.29) it is replaced back into the 11 original data values with the possibility of randomly being selected again. We repeat this 10 more times to get a new sample of 11 values. Our first resample ($k = 1$) or random selection may look like

$$13, \quad 9, \quad 9, \quad 13, \quad 10, \quad 10, \quad 9, \quad 9, \quad 9, \quad 9, \quad 10. \qquad (7.30)$$

This is the first bootstrap resample (again, $k = 1$). Note that the value 9 appeared four times in the original sample, but appears six times in the first resample. Also, the value 13 appeared once in the original sample, but 13 appears twice in the first resample. From this resample, one computes the median, which is 9, and the IQR of 9–10. Second, the whole process is repeated K times (where we will let $K = 20$ for our example). Thus, we generate 20 resample data sets ($k = 1, 2, 3, \ldots, 20$). From each of these sets, we compute the percentiles, that is, the median and 25th and 75th quartile. Next, we obtain the standard deviation of these percentiles samples). The process is simple. Note the results of the 20 bootstrap samples and the results are given in Tables 7.3 and 7.4, respectively.

Once again, this is a simple demonstration of the bootstrap technique on the NO_2 data. The technique is called a nonparametric bootstrap because we are not assuming

TABLE 7.3   20 Bootstrap Samples

Bootstrap Sample (NO$_2$) Data	25th Quartile	Median	75th Quartile
1	9	9	11
2	9	9	11
3	9	9	13
4	10	10	13
5	9	9	11
6	9	9	11
7	9	9	11
8	9	9	10
9	9	9	13
10	9	9	11
11	10	10	13
12	9	9	11
13	9	9	11
14	9	9	11
15	9	9	14
16	10	10	14
17	9	9	9
18	9	9	11
19	9	9	11
20	9	9	14

TABLE 7.4   Uncertainty Results from the 20 Bootstrap Samples

Statistic	Median Value	Standard Deviation	Coverage Factor 2
25th Quartile	9	0.366	(8.98, 9.32)
Median	10	1.268	(9.56, 10.74)
75th Quartile	11	1.417	(11.04, 12.36)

that the underlying process that generated the data follows a parametric distribution such as normality. This is a very handy technique when dealing with laboratory data that one cannot assume data is normally distributed. The only assumption we make is that the data is truly representative of the experimental process being analyzed. Also, as noted earlier, although our demonstration here was with a rather small sample, we usually need a moderately large sample for the bootstrap procedure to work well.

7.5   CALIBRATION BIAS

There are many ways of expressing bias in statistics, but it is basically the difference between what is recorded and what is actually expected. For example, if X_m is the mean of the sample and μ is the actual expected or "true" mean from the population, then we say that the bias is $X_m - \mu$. Suppose we have many labs participating in a study and using the same calibration technique and instrument for clinical chemistries. The expected resultant mean hematocrit value from all these laboratories for a particular trial sample is, say, 35.4. Suppose laboratory A derives a mean value of 34.9 for this same sample. The bias attributed to this one laboratory may be expressed as $|34.9 - 35.4| = 0.5$. Note that calibration reduces an error to an accepted range. The error is a precision error if it is estimated statistically. Otherwise, it is a biased error. The reader is referred to Beasley and Figliola (2000) for more details. Laboratory results can be precise but inaccurate. Inaccuracy can mean that there is bias. Bias is defined from a simple population perspective. Bias and calibration bias are examined by revisiting the method comparison procedures discussed earlier. Note that some authors discuss procedures for examining calibration bias. Some authors use the term validation of the calibration model. For our purposes, we will call it examining calibration bias. For a good discussion on the terms bias, calibration, calibration bias, calibration model, and other terms used in the literature, the reader is referred to Bias versus Calibration: http://elsmar.com/Forums/showthread .php?t=7404.

Now, let us refer to the method comparison of two instruments where one is the standard. A way to examine the bias of a new instrument versus the standard one is to plot the differences between the readings from each instrument versus the average of those readings. We refer one again to Vannest et al. (2002). Recall the section on method comparison in Chapter 3. We also discuss bias when comparing two instruments. This is precisely what we present here when evaluating Instrument A versus the standard. Figures 7.5–7.7 are based on simulated data to illustrate relative biases between the standard and Instrument A for measuring analytes, Albumin BCG, AST, and Phosphorus.

Note that in each of the figures how the plotted values all fall between the limits of -10 and $+10$ on the difference scale (vertical axis), which translated into the two instrument types or platforms being within $\pm10\%$ of each other. Thus, one concludes that the relative bias analysis of results between the test instrument and the standard demonstrated commutability of results between the two platforms.

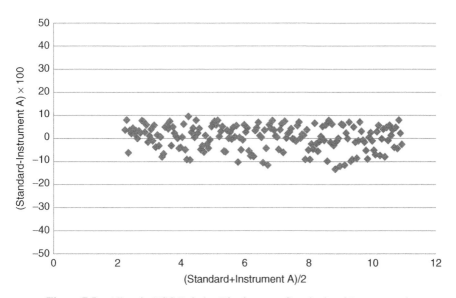

Figure 7.5   Albumin BCG Relative Bias between Standard and Instrument A

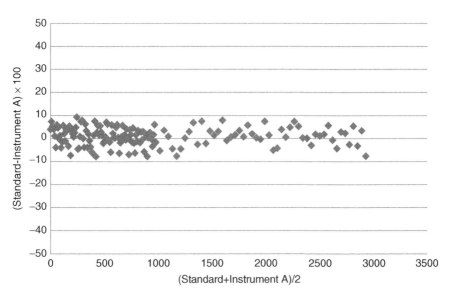

Figure 7.6   AST Relative Bias between Standard and Instrument A

Now, we discuss calibration bias in the most popular linear calibration model (Lavagnini and Magno 2007). We illustrate the calibration bias in the model with two examples. These examples illustrate calibration bias and no calibration bias, respectively. The linear calibration model is expressed as

$$y = b_0 + b_1 x, \tag{7.31}$$

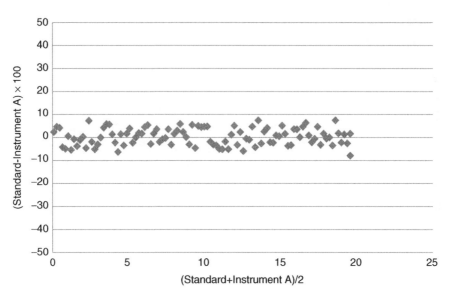

Figure 7.7   Phosphorus Relative Bias between Standard and Instrument A

where y, x, b_0, and b_1 are the instrumental response, concentration value, intercept, and slope, respectively, as discussed in this and other chapters.

There are approaches to evaluating the calibration bias in the calibration model (7.31). We will only discuss three simple approaches for the reader. Note that the implementation of and interpretations of results from fitting the linear calibration model are much easier as compared to curvilinear or nonlinear models (Ratkowsky 1990). If equivalent results can be achieved, the linear calibration models are preferred over curvilinear or nonlinear calibration models (Van Loco et al. 2002). Also, analytical methods usually work well on a calibration range if the response is essentially a linear function of the concentration of the analyte (Massart et al. 1997). Therefore, for keeping the basic concept of the calibration bias simple and straightforward, we will discuss linearity and measuring linearity. The first approach to examine linearity is the correlation coefficient (r) between x and y in the calibration model (7.31). Note that $r \geq 0.995$ is a good indicator of linearity but questions have been raised about the use of this indicator for linearity of the fitted linear model. The reader is referred to Massart et al. (1997) and Davies (1990) for more details and other statistical tests for the goodness of fit. One sometimes notes that $r \geq 0.995$ may be a good indicator of the calibration model to be nearly linear.

The second approach to evaluate the calibration bias in the model (7.31) is to examine the residual errors. As we recall from Chapter 3, a residual error is the difference between the observed and the corresponding predicted value from the fitted calibration model. We check a residual error plot for the underlying assumptions of normality of the residual errors and check for homoscedasticity or constancy of variance of the residuals. Thus, for evaluating the goodness of fit of the calibration model, the residual error plot is preferred over other tests. Note, for example, that in the residual error plot, the residual errors should be randomly distributed about an average residual error of zero; that is, no apparent trend toward either smaller or larger residual errors versus concentration.

The third approach to evaluate the calibration bias in the model (7.31) is to use the response factor (RF) or the relative standard deviation (RSD or %RSD) of the RF. Note that the RF (GC/MS Methods) is the ratio of the response of the instrument to the concentration of analyte. RSD is the absolute value of the coefficient of variation and an excellent indicator to measure linearity of calibration. Note that an analytical instrument can be said to be calibrated in any instance in which an instrumental response can be related to a single concentration of an analyte. Thus, linearity is determined by calculating the RSD for each analyte and comparing this RSD to the limit specified in the method. If the RSD does not exceed the specification, then linearity is assumed. Note that when the given criterion is met, the calibration is sufficiently linear to permit the laboratory to use an average RF, and it is assumed that the calibration is a straight line that passes through the zero/zero calibration point, that is, $(x, y) = (0, 0)$. We will demonstrate this. The EPA 600- and 1600-series methods (http://www.epa.gov/osw/hazard/testmethods/resources.htm) specify some criteria for determining linearity of calibration. We briefly summarize some upper bound numbers of linearity of calibration given by the EPA 600- and 1600-series methods. In the EPA 600- and 1600-series, the linearity specification varies from method to method, depending on the quantitation technique. For example, the typical limits on the %RSD are as follows:

1. 15% for the gas chromatography (GC) and high-performance liquid chromatography (HPLC) methods
2. 20% for analytes determined by the internal standard technique in the gas chromatography/mass spectrometry (GC/MS) methods (625 and 1625)
3. 20% for analytes determined by isotope dilution in Method 1625 and
4. 15% for mercury determined by atomic fluorescence. Note that metals methods that employ a linear regression specify a criterion for the correlation coefficient, r, such as 0.995.

7.5.1 Gas Chromatographic/Mass Spectrometric (GC–MS) Calibration Bias

In this section, we illustrate linearity of calibration models for GC/MS systems using three approaches discussed earlier, that is, the correlation approach, residual error plot approach, and RSD approach. Many GC/MS systems use calibrations and an internal standard quantification technique to test accuracy and document consistency over a range of concentrations. The recommended references include Method 625sBase/Neutrals and Acids (July 1991 Revision); Method 8270, Semivolatile Organic Compounds by Gas Chromatography/Mass Spectrometry (GC/MS) (1986), Method 525, Determination of Organic Compounds in Drinking Water by Liquid–Solid Extraction and Capillary Column Gas Chromatography/Mass Spectroscopy (1988), and Grob (1994).

Certain compounds are used as internal standards to track chemical behavior and compensate for sensitivity drift. Three or more levels of concentration of analytical standards are analyzed to establish a range of concentration and illustrate the linearity (first order) or curve linearity (second order) of the entire calibration curve. For a nonsignificant problem of calibration bias, we expect to have minimum %CV or the percent RSD over narrow calibration ranges. Let's first define terms used in GC/MS systems to calculate an RF: A_a = response of the analyte peak (area), A_{is} = response of the internal standard (area), C_a = concentration of the analyte, and C_{is} = concentration of the internal standard. Then the response factor (RF) for an analyte is defined as $RF = (A_a/A_{is})/(C_a/C_{is}) = (A_a C_{is})/(A_{is} C_a)$. Note that C_{is} is constant and A_{is} is shown to be essentially constant. Therefore, with GC/MS systems and in particular over narrow ranges of calibration we expect A_a to be directly proportional to C_a (Troost and Olavesen, 1996). That is, RFs are constant. Now we will examine the calibration bias in the calibration model (7.31) using the three approaches described earlier.

7.5.1.1 Simulated Example I Let's discuss the simulated data in Table 7.5 for linearity of the calibration model. We restrict analysis to means. By using the correlation approach, we check the correlation for linearity. The correlation of 0.976 between the mean A_a/A_{is} and mean C_a/C_{is} is less than 0.995, which implies that the calibration line is not linear (Figure 7.8a). Note the more appropriate curved fit to the five points. By using the residual error plot approach, residual errors in Figure 7.8b are not randomly distributed about an average residual error of zero, but this is difficult to see from only five data points. For simplicity, we will use %RSD for %CV. From Table 7.5, the %RSD of five mean RFs is 36.51, which is more than 20.00, implying a problem of calibration bias with the instrument. Troost and Olavesen (1996) recommended using all of the data points instead of means. By using all 30 data points, the correlation between A_a/A_{is} and C_a/C_{is} is 0.939, which is less than 0.995, raising the question of linearity in the calibration line (Figure 7.8c). The residual errors appear to be randomly distributed (Figure 7.8d) but the %RSD of all RFs, 44.79, is larger than 20.00, pointing to the calibration problem with the instrument. The same is true for the linear model in Figure 7.8d. The residual variance is about 0.10, denoting a wide spread of the points about the line. This fact becomes important when we examine our next data example, which satisfies all criteria for accurate calibration. Thus, in this case, two of the three approaches are consistent in concluding that the calibration line is not linear. That is, the instrument has a calibration problem.

7.5.1.2 Simulated Example II Now, let's discuss the simulated data in Table 7.6 for the linearity of the calibration model. By using the means, the correlation of 0.999 is larger than 0.995, suggesting that the calibration line is linear. By examining the plot of averages of A_a/A_{is} and C_a/C_{is} in Figure 7.9a, linearity appears reasonable. The scatter of the residuals in Figure 7.9b appear random, but as earlier, this is difficult to see with only five data points. By using the RSD approach, the %RSD of 4.44 is smaller than 20, strongly indicating the linearity in the calibration model (7.31). Thus, it appears that all three approaches confirm the linearity in the calibration line. Now, let's use all data points as recommended by Troost and Olavesen (1996). The

CALIBRATION BIAS

TABLE 7.5 GC–MS Calibration Bias – One Instrument Y, One Compound, Standard

Level (ng/μL)	C_a/C_{is}	C_a/C_{is} Mean	A_a/A_{is}	A_a/A_{is} Mean	RF
30	0.98		0.86		1.16
	0.34		0.34		
	0.88	0.60	1.40	0.70	
	0.29		0.48		
	0.91		0.93		
	0.20		0.17		
60	1.73		1.64		0.63
	1.08		0.26		
	1.76	1.34	1.25	0.84	
	1.23		0.30		
	1.72		1.27		
	0.55		0.35		
90	2.17		1.18		0.52
	1.84		0.64		
	2.80	2.13	1.83	1.20	
	2.00		0.91		
	2.16		1.60		
	1.82		0.52		
130	4.02		2.44		0.59
	2.90		2.03		
	3.95	3.35	2.74	2.14	
	2.99		0.91		
	4.03		2.17		
	2.24		1.58		
170	4.67		2.85		0.63
	3.71		1.84		
	4.35	4.11	2.21	2.76	
	3.83		2.82		
	4.83		2.96		
	3.28		2.86		

%RSD of mean RFS = 36.51, %RSD of RFS using all data points = 44.79.

correlation of the A_a/A_{is} versus C_a/C_{is}, 0.9959 is larger than 0.995 (Figure 7.9c). The residual errors in Figure 7.9d are randomly distributed about an average residual error of zero. The residual variance for this model is 0.01, which is one-tenth of the residual variance for the previous model in Figure 7.8c and d. Note the much tighter fit of the points to the line of the model in Figure 7.9c as opposed to Figure 7.8c. Lastly for this data, by using the RSD approach, the %RSD of 8.44 is smaller than 20, strongly indicating the linearity in the calibration model (7.31). Again, all three approaches are consistent in confirming the linearity in the calibration line.

From the available literature and from two simulated examples, we conclude that the instrument has a problem of calibration using example 1 but no calibration

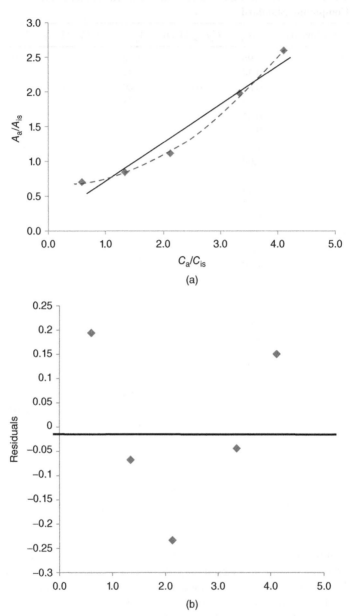

Figure 7.8 (a) GC–MS Calibration Bias – Average, One Instrument Y and One Compound.
(b) GC–MS Calibration Bias – Residual, One Instrument Y and One Compound. (c) GC–MS
Calibration Bias – All Data, One Instrument Y and One Compound. (d) GC–MS Calibration
Bias – Residual, One Instrument Y and One Compound

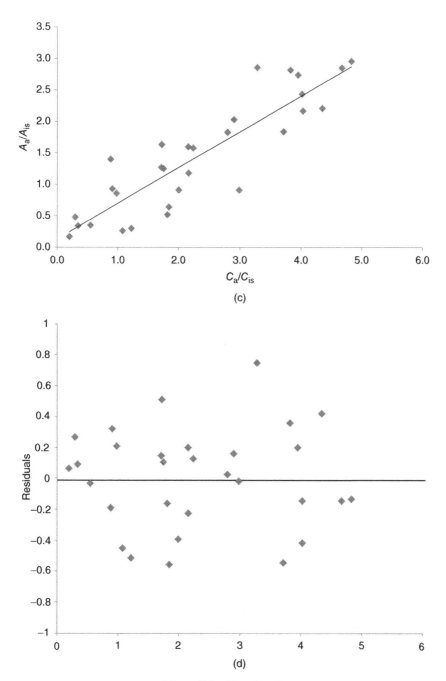

Figure 7.8 (*Continued*)

TABLE 7.6 GC–MS Calibration Bias – One Instrument Z, One Compound, Standard

Level (ng/μL)	C_a/C_{is}	C_a/C_{is} Mean	A_a/A_{is}	A_a/A_{is} Mean	RF
20	1.13	0.70	0.68	0.64	0.91
	0.40		0.31		
	0.83		0.88		
	0.36		0.37		
	1.34		1.35		
	0.14		0.22		
50	1.75	1.29	1.32	0.95	0.83
	0.97		1.06		
	1.65		1.10		
	0.66		0.45		
	1.78		1.64		
	0.92		0.87		
80	2.06	1.92	2.64	1.66	0.93
	1.79		1.26		
	2.22		2.42		
	1.46		0.99		
	2.58		1.88		
	1.41		1.57		
120	3.78	3.01	2.68	2.59	0.86
	2.39		1.63		
	3.72		3.20		
	2.18		2.34		
	3.06		3.36		
	2.94		2.32		
160	4.48	4.04	3.53	3.36	0.83
	2.96		2.65		
	4.86		3.72		
	3.87		3.13		
	4.91		4.08		
	3.14		3.05		

%RSD of mean RFS = 4.44, %RSD of RFS using all data points = 8.44.

problem in example 2. We have determined the linearity by calculating %RSD and comparing the value with the limit specified in the EPA method. Since the %RSD does not exceed the specification in the second example, the linearity is assumed. Thus, the calibration is sufficiently linear for the laboratory to use an average RF and assume that the calibration is a straight line that passes through the zero/zero calibration point. From the analyses of the examples given in Tables 7.5 and 7.6, the intercepts are not statistically different from 0 (t-value $= 0.2602$, p-value $= 0.7967$ and t-value $= 1.1248$, p-value $= 0.2702$, respectively). Therefore, the results of these examples confirm the assumption of the calibration line passing through the zero/zero calibration point.

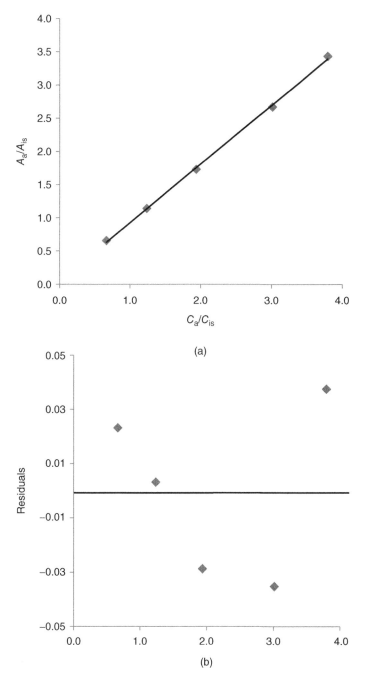

Figure 7.9   (a) GC–MS Calibration Bias – Average, One Instrument Y and One Compound. (b) GC–MS Calibration Bias – Residual, One Instrument Y and One Compound. (c) GC–MS Calibration Bias – All Data, One Instrument Y and One Compound. (d) GC–MS Calibration Bias – Residual, One Instrument Y and One Compound

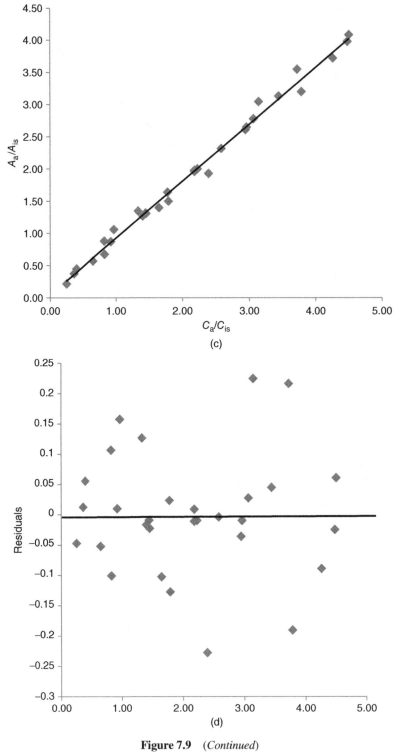

Figure 7.9 (*Continued*)

7.5.2 Discussion

If the calibration is not linear, another mathematical function should be employed to relate the instrument response to the concentration. However, properly maintained and operated lab instrumentation should have no difficulty in meeting linearity specifications for the EPA 600- and 1600-series methods. Linear regression emphasizes the importance of higher concentration standards and that the correlation coefficient is impacted little by poor performance of calibration standards with low concentrations. Thus, most analytical methods focus on a calibration range where the response is essentially a linear function of the concentration of the analyte.

7.6 MULTIPLE INSTRUMENTS

Suppose we have the following setup: multiple instruments X, Y, and Z, and compounds A, B, C, D, and E (Table 7.7). If there are no calibration problems, two (primary) things should happen: (1) Within any one instrument for any one compound a plot of A_a/A_{is} versus C_a/C_{is} should be linear, (2) There should be no difference in the %RSDs, that is, %CVs among the three instruments. In Section 7.5, we discussed calibration bias problems with an instrument based on plots of A_a/A_{is} versus C_a/C_{is}, residual errors; response factors (RFs), and %RSDs. In this section, we discuss differences in the %RSDs (%CVs) of the three instruments X, Y, and Z. Recall for accurate calibration there should be no difference in the %RSDs among the three instruments, X, Y, and Z.

By looking at Figure 7.10, there appear to be differences in the mean %RSD readings of the three instruments X, Y, and Z. In addition, the outlier instrument appears to be instrument Y. Let's test if the three mean %RSDs are the same using a one-way analysis of variance (ANOVA) statistical procedure (see Chapter 2, Section 2.3.4). Note across all three instruments comparing the mean %RSD readings, we have the ANOVA results in Table 7.8. The calculated value of F is 3.8852 with 2 and 12 between groups and within group degrees of freedom, respectively, yielding an area under the F-curve or p-value of 0.000023. Therefore, the overall test is statistically significant (<0.05) rejecting the equality of mean %RSD readings for the three instruments X, Y, and Z. The reader is referred to Chapter 2 for more details regarding using ANOVA results for testing the equality of means.

Now, since overall significance is found statistically, we need to conduct pairwise comparisons (X vs. Y, X vs. Z, or Y vs. Z) to determine where the difference occurs to answer the question of the instrument Y calibration and/or performance under question. This is called a multiple comparison procedure and requires an adjustment to the p-value.

Now, we use the multiple comparisons procedure called the Tukey–Kramer (Tukey 1953; Kramer 1956) test or the honestly significant difference (HSD) test. The procedure is detailed in Chapter 2. We use the expression (2.30) in Chapter 2 for the HSD test, where α is the level of significance (0.05), k is the number of means in the experiment, N is the total sample size, and $N - k$ is called the error degrees of freedom

TABLE 7.7   Response Factor Standard Level Concentration

Compound	10	30	60	100	%RSD (%CV)
		Instrument X			
A	1.130	1.203	1.058	1.175	5.54
B	1.120	1.171	1.078	1.161	3.75
C	1.115	1.220	1.079	1.182	5.55
D	1.110	1.091	1.188	1.220	5.35
E	1.213	1.125	1.170	1.060	5.73
		Instrument Y			
A	0.900	1.120	1.300	1.150	15.04
B	0.951	1.225	1.179	1.272	12.30
C	0.900	1.423	1.279	1.298	18.44
D	0.920	1.323	1.369	1.398	9.30
E	0.961	1.120	1.179	1.198	10.23
		Instrument Z			
A	1.095	1.203	1.062	1.167	5.71
B	1.122	1.203	1.076	1.156	4.71
C	1.074	1.177	1.071	1.172	5.25
D	1.130	1.035	1.131	1.201	6.06
E	1.065	1.165	1.202	1.160	5.09

TABLE 7.8   ANOVA-Single Factor-Summary Results

SUMMARY

Groups	Count	Sum	Average	Variance
X	5	25.9223	5.18447	0.6577
Y	5	73.0308	14.6061	13.7992
Z	5	26.8195	5.36391	0.2809

ANOVA

Source of Variation	SS	df	MS	F	p-value	F crit
Between Groups	290.3670	2	145.1835	29.5531	2.31E-05	3.8852
Within Groups	58.9514	12	4.9126			
Total	349.3184	14				

(Error df). The quantity, q, is obtained from published statistical tables for the values of α, k, and $N-k$ (Pearson and Hartley 1958). The value, MSE, is the mean-squared error or the within groups means square (MS) as seen for example in Table 7.8. The values, n_i and n_j, are the sample sizes of the individual groups i and j where, in our aforementioned case, i and j can take on the labels of X, Y, or Z, since we have three instruments as X, Y, and Z. Note in Table 7.8, the average readings of %RSDs for

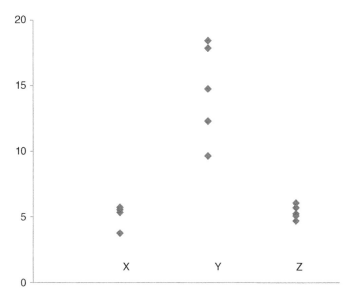

Figure 7.10   Means %RSDs for the Three Instruments X, Y, and Z

instruments X, Y, and Z are 5.18, 14.61, and 5.36, respectively. Thus, the absolute difference between the mean %RSD readings is |5.18 − 14.61| = 9.43. The MSE from Table 7.8 is 4.9126. $N = 15$, $k = 3$, $N − k = 12$, $n_X = 5$ and $n_Y = 5$. By referring to Pearson and Hartley (1958), we have the value of $q_{\alpha, k, N-k} = q_{0.05, 5, 12} = 3.77$. Thus, for means X and Y the HSD test value in the expression (7.32) is as follows:

$$HSD = 3.77 \sqrt{\frac{4.9126}{2} \left(\frac{1}{5} + \frac{1}{5} \right)}$$

$$= 3.74. \tag{7.32}$$

Since the absolute difference of 9.43 between X and Y is greater than 3.74, two mean %RSD readings are statistically different. We perform similar comparisons for X versus Z and Y versus Z in Table 7.9.

Note that the Tukey–Kramer test or the HSD test is consistent with the overall rejection of the null hypothesis of the equality of the three mean %RSD readings and

TABLE 7.9   Absolute Mean Differences and HSD Pairwise Comparisons

Comparison	Absolute Mean Difference	HSD	Means Statistically Different?
X versus Y	9.43	3.74	Yes
X versus Z	0.18	3.74	No
Y versus Z	9.25	3.74	Yes

answers the question of the instrument Y calibration and/or performance compared to X and Z under question.

Since we have only five observations per instrument, one could argue that the sample size is too small for an ANOVA comparison that assumes normality. Thus, we can use the nonparametric Wilcoxon comparison as discussed in Chapter 2. When we apply this method the overall p-value comparing the mean %RSDs of the three instruments is $p = 0.0092$, which is consistent with the ANOVA result, in that we reject the null hypothesis of the equality of the three means. Comparing the %RSD of the instruments pairwise, we have X versus Y, $p = 0.0122$, Z versus Y, $p = 0.0122$ and X versus Z, $p = 0.9168$. Thus, once again, instrument Y is the significant outlier.

7.7   CRUDE VERSUS PRECISE METHODOLOGIES

In a calibration problem, suppose we have an inexpensive, convenient, less precise measurement technique called a crude technique, which we label as C. Let's also suppose we have an expensive, inconvenient, highly precise technique, which we label, P. Cost considerations and convenience make the crude technique a good trade-off and thus, more attractive despite its less precision. The goal of the calibration problem is to use the value of the crude method to estimate what would have been obtained from the precise method. This is precisely a regression problem. Now, we detail the technique. Suppose in an initial experiment we have data from both the precise technique and the crude technique. We compute the equation

$$C = b_0 + b_1 \ P, \tag{7.33}$$

where b_0 and b_1 are the intercept and slope, respectively, as we've discussed in this and other chapters.

As we recall from this and other chapters, equation (7.33) is a simple linear equation because it has one independent variable P. Note that in the calibration problem, the error is associated with the predicted crude (C), while we have no error associated with precise (P). Because of this reason, the use of inverse regression is recommended in the literature (Neter, Wasserman and Kutner 1989, Section 5.6). In the inverse regression, the crude technique is regressed on the precise technique using the estimates b_0 and b_1 from the initial experiment. Thus, in future experiments, we can collect data on C and for convenience use equation (7.34) to estimate the precise value of P, that is,

$$P = \frac{(C - b_0)}{b_1}. \tag{7.34}$$

As an example, a particulate matter (PM) data set is given in Table 7.10. Note that we have the data in the first column of Table (7.10) labeled as Tech_P, and the data in the second column labeled as Tech_C. The data is usual PM data in $\mu g/m^3$, which is all below a standard set at $12 \, \mu g/m^3$.

TABLE 7.10 Simulated Data on Tech_P and Tech_C

Example (Initial Experiment)	
Tech_P	Tech_C
1.3	1.5
2.6	2.5
3.4	3.7
5.7	5.2
7.3	6.9
8.4	8.5
9.7	9.6

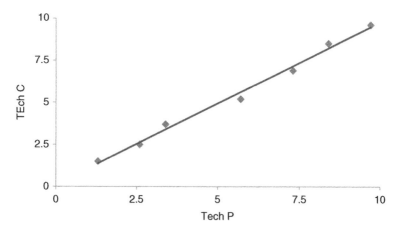

Figure 7.11   Fitted Regressions between Tech C and Tech P

Now, we regress Tech_C on Tech_P by using equation (7.34). This yields the equation

$$Tech\_C = 0.1247 + 0.9643 \ Tech\_P. \tag{7.35}$$

Thus, from the fitted simple linear regression (7.35), we have the estimates of intercept and slope: $b_0 = 0.1247$ and $b_1 = 0.9643$. Note that the estimated slope is positive. From the plot of Tech P versus Tech C in Figure 7.11, the regression fit is appropriate ($R^2 = 0.9918$). Therefore, there is a significantly strong positive association between P and C with a p-value of 0.0001. The regression summary statistics are given in Table 7.11.

Referring to Chapter 3, the intercept is not significantly different from zero and the slope is not significantly different from one, leading one to conclude that the two methods are comparable. If the calibration is correct for either method or machine, we can clearly use C (more convenient, less expensive, etc.) to predict P by using

TABLE 7.11   Regression Summary Statistics

	Coefficient Value	Standard Error	t-Statistic	p-value
Intercept	0.1247	0.243	0.512	0.630
Tech P	0.9643	0.039	24.638	<0.0001

equation (7.36), that is,

$$P = (C - b_0)/b_1$$
$$= (C - 0.1247)/0.9643. \tag{7.36}$$

Suppose in a future experiment we just have crude values, C: 1.42, 2.46, 3.70, 5.26, 7.03, 8.68, and, 9.82. Then from equation (7.36) we can easily predict precise P (predicted) values given as follows: 1.34, 2.42, 3.71, 5.33, 7.16, 8.87, and 10.05.

Note that in the statistical literature the method of inverse regression is known as the classical method (Neter, Wasserman and Kutner 1989, Section 5.6), whereas in the calibration literature the method of regressing P on C directly is called the inverse method (Chow and Shao 1990). For more details, the reader is referred to the online discussion by Dallal (2000) and the Center for Professional Innovation and Education, Inc, (CfPIE©) http://www.cfpie.com/showitem.aspx?productid=046. As discussed by Dallal (2000), "Calibration and comparability differ in one important respect. In the comparability problem, both methods have about the same amount of error (reproducibility). Neither method is inherently more accurate than the other." In the calibration problem, an inexpensive, convenient, less precise measurement technique (labeled C, for "crude") is compared to an expensive, inconvenient, highly precise technique (labeled P, for "precise"). For considerations of cost and convenience, the use of the inverse method to predict P makes the crude technique a good trade-off despite its lack of precision.

REFERENCES

A Resampling Method Called the Bootstrap. Accessed July 22, 2015 from http://warnercnr .colostate.edu/~gwhite/fw663/bootstrap.pdf.

ANSI/NCSL Z540-2 (1997). Guide to the Estimation of Uncertainty in Measurement. (GUM).

Barr D and Zehna P. (1983). Probability: Modeling Uncertainty, Addison-Wesley, Boston, MA.

Beasley DE and Figliola RS. (2000). Theory and Design for Mechanical Measurements 3rd ed., John Wiley &Sons, Hoboken, NJ.

Bias vs. calibration. http://elsmar.com/Forums/showthread.php?t=7404.

Birch K. (2001). British Measurement and Testing Association, Measurement Good Practice Guide No.36 1, Middlesex, United Kingdom, TW11 0NQ.

Center for Professional Innovation and Education, Inc, (CfPIE©). Accessed July 22, 2015 from http://www.cfpie.com/showitem.aspx?productid=046.

Chow S and Shao J. (1990). On the difference between the classical and inverse methods of calibration. Journal Royal Statistical Society, Series C 39: 219–228.

Dallal GE. (2000). http://www.jerrydallal.com/lhsp/compare.htm).

Davies, NW. (1990). Gas chromatographic retention indices of monoterpenes and sesquiterpenes on methyl silicone and Carbowax 20M phases. Journal of Chromatography 503: 1–24.

Efron B. and Tibshirani RJ. (1993). An Introduction to the Bootstrap. Monographs on Statistics and Applied Probability. No. 57. Chapman and Hall, London, 436.

Environmental Protection Agency (EPA). (1997). Guide to Uncertainty and Measurement.

EURACHEM/CITAC Guide CG-4 (2000). Quantifying Uncertainty in Analytical Measurement, 2nd ed., ANSZ-ASQ, National Accreditation Board.

Grob K Jr. (1994). Injection techniques in capillary GC. Analytical Chemistry 66 (20): 1009A–1019A.

Kramer CY. (1956). Extension of multiple range tests to group means with unequal sample sizes. Biometrics 63: 307–310.

Lavagnini I, and Magno F. (2007). A statistical overview on univariate calibration, inverse regression, and detection limits: application to gas chromatography/mass spectrometry technique. Mass Spectrometry Reviews 26(1):1–18.

Massart DL, Vandeginste BGM, Buydens LMC, De Jong S, Lewi PJ and Smeyers-Verbeke, J. (1997). Handbook of Chemometrics and Qualimetrics: Part A, Elsevier, Amsterdam.

Method 625sBase/Neutrals and Acids. (July 1991 Revision.). Code of Federal Regulations, Part 136; Title 40; Office of the Federal Register, National Archives and Records. Administration, U.S. Government Printing Office: Washington, DC, Appendix A.

Method 8270, Semivolatile Organic Compounds by Gas Chromatography/Mass Spectrometry (GC/MS): Capillary Column Technique.

Method 525, Determination of Organic Compounds in Drinking Water by Liquid–solid Extraction and Capillary Column Gas Chromatography/Mass Spectroscopy, Revision 2.1. Methods for the Determination of Organic Compounds in Drinking Water (1988); Environmental Monitoring Systems Laboratory, Office of Research and Development, U.S. EPA: Cincinnati, OH.

Method 1625, Revision B (991), Semivolatile Organic Compounds by Isotopic Dilution GC/MS. Code of Federal Regulations, Part 136; Title 40; Office of Federal Register, National Archives and Records Administration, U.S. Government Printing Office: Washington, DC; Appendix A.

Method 8270, Semivolatile Organic Compounds by Gas Chromatography/Mass Spectrometry (GC/MS) (1986).

Neter J, Wasserman W and Kutner MH. (1989). Applied Linear Regression Models, 4th ed. with Student CD, McGraw Hill/Irwin Series: Operations and Decision Sciences. Homewood, IL.

Pearson ES and Hartley HO. (1958). Biometrika Tables for Statisticians, 2nd ed., Vol. I, Cambridge University Press, New York.

Ratkowsky DA. (1990). Handbook of Nonlinear Regression Models. Marcel Dekker: New York.

Reda E. (2011). Method to Calculate Uncertainties in Measuring Shortwave Solar Irradiance Using Thermopile and Semiconductor Solar Radiometers. Technical Report NREL/TP-3B10-52194, National Renewable Energy Laboratory, July.

Taylor BN and Kuyatt C, (1994). Guidelines for Evaluating and Expressing the Uncertainty of NIST Measurement Results, NIST Technical Note 1297, National Institute of Standards and Science, Gaithersburg, Maryland, USA.

Troost JR and Olavesen EY. (1996). Gas chromatographic/mass spectrometric calibration bias. Annals of Chemistry 68:708–711.

Tukey JW. (1953). The Problem of Multiple Comparisons. Mimeographed Monograph. Princeton University. cited in Y. Hochberg and AC. Tamhane. Multiple Comparison Procedures. John Wiley and Sons. New York.

Van Loco J, Elskens M, Croux C and Beernaert H. (2002). Linearity of Calibration Curves: Use and Misuse of the Correlation Coefficient. Accreditation and Quality Assurance 7: 281–285.

Vannest R, Douglas J, Markle V and Bozimowski D. (2002). Evaluation of Chemistry Assays on the Abbott ARCHITECT® c8000™ Clinical Chemistry System. The Paper Presented at the 54th Annual Meeting of the American Association of Clinical Chemistry, July 28-August 1, 2002. *Abbott Laboratories*: Irving TX.

Wastes – Hazardous Waste – Test Methods: Resources, U.S. Environmental Protection Agency, http://www.epa.gov/osw/hazard/testmethods/resources.htm.

8

ROBUSTNESS AND RUGGEDNESS

8.1 INTRODUCTION

In this chapter, we want to discuss the two methodologies of robustness and ruggedness. Some authors use the terms interchangeably. They deal with being able to reproduce an analytical method in different laboratories or under different circumstances (inputs) without changing the results. This is somewhat oversimplified, as there are subtle differences between the two approaches. We discuss the focus of both methods with applications. For a further discussion with examples, see Vander Heyden et al. (2001) and the ICH (International Conference on Harmonisation) Harmonised Tripartite Guideline (1994, 1996).

The most widely used application of robustness testing is best known in the pharmaceutical industry because of the strict validation requirements and regulations in that area set by regulatory authorities. Thus, most of the ICH definitions and existing methodologies can be found in that domain. Similar to most statistical methods, it has wide implications and robustness testing of analytical methods, as we shall discuss, can be applied in other fields.

Robustness is the characteristic of the process output or response to be insensitive to the variation of the inputs. Setting the process targets using the process interactions increases the likelihood of the process exhibiting robustness. Robustness has traditionally not been considered a validation parameter in the strictest sense. It is usually investigated during method development once the method is at least partially optimized.

Vander Heyden et al. (2006) have developed an excellent tutorial of the concept of robustness. They note that the robustness test was considered a part of method

Introduction to Statistical Analysis of Laboratory Data, First Edition.
Alfred A. Bartolucci, Karan P. Singh, and Sejong Bae.
© 2016 John Wiley & Sons, Inc. Published 2016 by John Wiley & Sons, Inc.

validation related to the precision (reproducibility) determination of the method. They also note that performing a robustness test late in the validation procedure involves the risk that when a method is found not to be robust, it should be redeveloped and optimized. At this stage, much time, effort, and money have already been invested in the optimization and validation, and therefore, one wants to avoid this.

The steps in a test of robustness and ruggedness are much the same and fairly standard. They follow the steps one might anticipate in a regular design and analysis of an experiment. They also follow the steps listed by Vander Heyden et al. (2006) and are as follows:

1. Identify factors to be tested.
2. Establish the different levels for the factors.
3. Select the appropriate experimental design.
4. Define the experimental protocol (complete experimental setup).
5. Define the responses to be determined and the statistical analyses to be performed.
6. Perform the experiments and determine the responses.
7. Calculate the effects.
8. Derive the statistical and/or graphical analysis of the results.
9. Generate an appropriate report of the findings from the statistical analysis.

8.2 ROBUSTNESS

For our purposes, we'll keep the procedure of robustness to a simple example. We assume during a robustness study that one usually intentionally varies the method parameters (internal influences, factors) to see if the method results are affected. In liquid chromatography (LC) some examples of typical variation are as follows:

• Number, type, proportion of organic solvents
• Buffer composition and concentration
• Temperature
• Flow rate
• Gradient variations
• Hold/storage times

Many experiments have been done considering IL-8 mRNA expression (see John et al. 1998) for various applications. We consider an example in a gene expression system to confirm the robustness of that system. The factors we consider are intervention (treatment and control), relative luminescence (RLU) conditions (frozen/thaw and hold times in weeks) and the method result is mRNA levels of IL-8 and TNF-α. The experimental setup is as follows: Cells were treated with either vehicle (control)

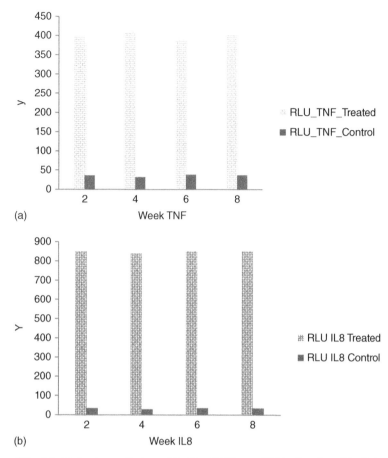

Figure 8.1 (a) Robustness Test – Weekly TNF-α mRNA Levels. (b) Robustness Test – Weekly IL-8 mRNA Levels

or chemical toxicant (treated) to induce the TNF-α and IL-8 genes. The cells were lysed (cell death) and the lysates frozen at $-80\,°C$. They were thawed at weekly intervals and the IL-8 and TNF-α mRNA levels were measured in quadruplicate over an 8-week period.

The results of the experiment are given in Figure 8.1a and b, which are bar charts of the TNF-α and IL-8 levels, respectively, for the treated and control groups. The percent coefficient of variations for replicates performed on a single day within a day (intra) and across weeks (inter) are given in Table 8.1. Note that over the 8-week period the %CV ranged from 0.48% to 4.65% for assay replicates performed on a single day (intra) of the week. The %CV ranged from 0.59% to 8.64% for assays performed over the 8 weeks (inter).

Conclusion: This instrument is considered robust for measuring mRNA levels since the assays have the ability to maintain high intra-experimental and interexperimental precision over a wide range of RLU values.

TABLE 8.1   Percent CV for Robustness Study

Treatment Group	Intra %CV	Inter %CV
TNF-α treated	1.23	2.07
TNF-α control	3.76	7.35
IL-8 treated	0.48	0.59
IL-8 control	4.65	8.64

So in summary, why do we have robustness? Recall: During our robustness application, one usually intentionally varies the method parameters to see if the method results are affected. What parameters (internal influences) were varied? They were treatment (control, treated) and RLU conditions (frozen/thaw and hold times (weeks)).

What were the method results? These obviously were the mRNA values.

8.3   RUGGEDNESS

The USP (United States Pharmacopeia 2007) defines ruggedness as "the degree of reproducibility of test results obtained by the analysis of the same samples under a variety of normal test conditions such as:

- Different laboratories
- Different analysts
- Different instruments
- Different reagent lots
- Different analysis days
- Different elapsed assay times
- Different assay temperatures

Ruggedness measures the lack of external influences on the test results, "external" meaning they are not well controlled. The purpose of ruggedness is to determine the variables (external experimental factors) that strongly influence the measurements provided by the method and to determine how closely these variables need to be controlled. If one were to discover that examining reasonable ranges of factors in an experimental method that they do not strongly (statistically) influence the measurements or results, then we say that the method is rugged for the factors over the ranges tested. We will explain and demonstrate this approach. Ruggedness tests do not determine the optimal conditions for the test method. Ruggedness tests should precede an interlaboratory study, that is, they should initially be done within a single laboratory. The interlaboratory ruggedness study then should be done using the other laboratory as an additional factor, and this usually leads to determining the precision of the test method. As opposed to the reproducibility definition using the coefficient of variation as we did for robustness, here we concentrate on factors causing variability in the system. Thus, an objective of a ruggedness test is the evaluation of potential factors that may cause variability in the assay responses of the method, for example, content determinations such as temperature and time on determining the Furan content in foods (Becalski and Seaman 2005). For this purpose, small variations in the method parameters may exist. The examined factor intervals, of course, must be representative of the variations expected if one were to transfer the method between laboratories or instruments. The factors are then usually studied in an experimental design approach.

Performing a test of ruggedness is not difficult, but it is mathematically and statistically cumbersome. Therefore, we are going to introduce and summarize the procedure using the guidelines of the American Society of Testing and Methods (ASTM), Standard Guide for Conducting Ruggedness Tests (E1169-02). The entire guide is available to order on the web.

We perform a ruggedness test as follows:

We suppose our experiment will be an HPLC (High-Performance Pressure Liquid Chromatography) assay. Suppose also that we have seven experimental factors of interest. Our goal is to determine which of these factors really influence or cause undue variation in the outcome of the experiment and may have to be controlled more carefully. For the sake of demonstration purposes only, our outcome could be anything such as percent main compound (%MC) as in Vander Heyden et al. (2006) or other value. To set up our ruggedness experiment, we follow steps (1)–(9) in Section 8.1. We go through the 9 steps.

1. **Identify the factors to be tested.**
   The factors to be included in a ruggedness test are related to the analytical procedure (operational or experimental factors) and to the environmental conditions (environmental factors). The operational factors are selected from the description of the analytical method (i.e., laboratory procedures and equipment needed to perform the experiment), whereas the environmental factors are not necessarily specified explicitly in the analytical method per se. For our purposes, the selected factors can be quantitative (continuous, sometimes referred to as interval), qualitative (discrete or categorical), or mixture factors. Table 8.2 indicates a list of factors that could be considered during ruggedness testing in our HPLC experiment. The list is not exhaustive but gives an idea of some factors commonly examined. Also, in Table 8.2, there is a level given for each factor. The continuous factors are % solvent, column temperature, buffer concentration, and pH of buffer. These factors all have a numeric assignment. The categorical or qualitative factors are analyst, column manufacturer, and reagent, which are expressed as a nominal level. Note that each factor is labeled from A to G.
   We do not have mixture factors in our design. However, we should explain what they are. The properties of a mixture are almost always a function of the relative proportions of the inputs rather than their absolute amounts. In experiments with mixtures, a factor's value is its proportion in the mixture, which falls between zero and one. The sum of the proportions in any mixture factor is one (100%). Thus, with mixtures, it is impossible to vary one component of the factor independently of all the others. When you change the proportion of one component,

TABLE 8.2   A List of Factors That Could Be Considered During Ruggedness Testing in the HPLC Experiment

Factor	Label	Units	Level (−)	Level(+)
Analyst	A	—	Analyst 1	Analyst 2
Solvent	B	%	23	27
Column manufacturer	C	—	Manufacturer 1	Manufacturer 2
Column temperature	D	°C	30	40
Buffer concentration	E	%m/v[a]	0.230	0.250
pH of buffer	F	—	6.3	7.3
Reagent	G	—	Reagent 1	Reagent 2

[a]Mass of solute in g per 100 mL of total solution.

the proportion of one or more other components must also change to compensate so that their total sums to 100%. This simple fact has an important effect on every aspect of experimentation with mixtures, that is, the factors, the design properties, and the interpretation of the results. As a simple example of a mixture experiment (DOT 2012), consider concrete as a mixture of three components: water, cement, and aggregate, where each component represents the volume fraction of the factor, concrete. The volume fractions of the components sum to one or 100%.

2. **Establish the different levels for the factors.**
   Note the last two columns of Table 8.2. The levels of each factor are indicated as a plus (+) and minus (−). A (+) for a given factor indicates that the measurement is made with that factor set at the high level, and a (−) indicates the factor is to be at the low level. All seven factors are set at two levels. For example, if temperature is set at two levels, 30° and 40°. The 30 is assigned a "−" and the 40 is assigned a "+." This assignment applies in Table 8.3.

3. **Select the appropriate experimental design.**
   The way we proceed is that we run the experiment at different combinations of these factor levels and determine the outcome at each run. These designs generally follow a Plackett–Burman design. In 1946, Plackett and Burman published their pioneering paper "The Design of Optimal Multifactorial Experiments" in *Biometrika* (vol. 33). This paper described the construction of very economical designs with the run (replicated experiment) number as a multiple of four (rather than a power of 2, which were the most common designs). In other words, we can have designs of $N = 4, 8, 12, 16, 20$, and so on runs, each run having $N − 1$ factors. The reason for this will become clear when we explain the design pattern for this experiment. Most designs chosen are either 8 or 12 runs. Recall, each variable (factor) is represented at two levels, namely, "high" and "low," which define the upper and lower limits of the range covered by each variable. In addition to the variables of real interest, the Plackett–Burman design may consider insignificant dummy variables, whose number should be one-third of all variables. The dummy variables, which are not assigned any values (usually a +1 or −1 denoting the level), introduce some redundancy required by the

statistical procedure. For computational statistical reasons not covered in detail here, incorporation of the dummy variables into an experiment allows an estimation of the variance (experimental error) of an effect. A good example of the use of dummy variables can be seen in Jain et al. (2010) and Vander Heyden et al. (2006).

Plackett–Burman designs are called screening designs as they screen for the most important factors influencing any variation in the experimental outcome. They are considered to be very efficient and only main effects are of interest. That is to say, the main effects or factors are considered independent contributors to the outcome and do not necessarily interact with each other. In other words, the calculated effect of any factor on the outcome is not greatly affected by any changes in the levels of any of the other factors. Also, the levels within any factor do not differ too greatly from each other. There are other more complicated designs that determine this interaction effect of the factors. Returning to our situation, Table 8.3 is an example of an 8-run design conforming to the factors we considered in Table 8.2. Note that in Table 8.3 the factors are across the top and represent the columns in the table. The runs are on the left as are the rows in the table. This particular design sequence $(+,-)$ for eight runs across seven factors was first suggested by Yates (1935). There are many versions of the Plackett–Burman design. The important thing to remember is that each factor within a run (column) must have the same number of $+$'s and $-$'s. To construct such a design is easy once you determine the first row. There are certain guidelines to help with this (Vander Heyden and Massart 1996). As one example, refer to Table 8.3. This is a single experiment: $N = 8$ runs, $N - 1 = 7$ factors with response R1–R8. Note that the first row has 4 $+$'s and 3 $-$'s to represent four factors at their upper level and three factors at their lower level. This is just one possible first row in a seven-factor experiment with eight runs. The following $N - 2$ rows in Table 8.3 are obtained by a cyclical permutation of one position (i.e., shifting the line by one position to the right) compared to the previous row.

TABLE 8.3 Plackett–Burman Design Construction Pattern

Run	Ruggedness Factors							Results
	A	B	C	D	E	F	G	Exp. 1
1	+	+	−	+	+	−	−	R1
2	−	+	+	−	+	+	−	R2
3	−	−	+	+	−	+	+	R3
4	+	−	−	+	+	−	+	R4
5	+	+	−	−	+	+	−	R5
6	−	+	+	−	−	+	+	R6
7	+	−	+	+	−	−	+	R7
8	−	−	−	−	−	−	−	R8

TABLE 8.4   Plackett–Burman Design for Ruggedness Experiment

Run			Ruggedness Factors					Results (%MC)	
	A	B	C	D	E	F	G	Exp. 1	Exp. 2
1	−	−	−	−	−	−	−	96.8	96.5
2	−	−	+	−	+	+	+	100.5	100.7
3	−	+	−	+	−	+	+	100.2	99.7
4	−	+	+	+	+	−	−	98.7	97.8
5	+	−	−	+	+	−	+	99.9	99.5
6	+	−	+	+	−	+	−	101.8	101.7
7	+	+	−	−	+	+	−	101.6	101.5
8	+	+	+	−	−	−	+	98.3	98.1

This means that the sign of the first factor (A) in the second row is equal to that of the last factor (G) in the first row. The signs of the following $N-2$ factors in the second row are equal to those of the first $N-2$ factors of the first row. The third row is derived from the second one in an analogous way. This procedure is repeated $N-2$ $(8-2=6)$ times until all but one line is formed. The last (Nth) row consists only of minus signs.

A Plackett–Burman design with N experiments can examine up to $N-1$ factors. This procedure works for any Plackett–Burman design of N runs and $N-1$ factors. Table 8.4 by Yates (1935) is a variation on the construction pattern, but achieves the same results in design specifications such as each column (factor) having the same number of high levels and low levels within a factor.

4. **Define the experimental protocol (complete experimental setup).**
   Table 8.4 is the summary of the experimental setup. Each of the seven factors (columns) was included in each of eight runs (rows) at their various levels of effect $(-,+)$. It is assumed that each of the eight runs is performed under the same experimental conditions within the laboratory. These may be environmental conditions such as temperature and humidity. If the runs cannot be performed on the same day, then to avoid any time effect, the laboratory conditions should be the same or similar on the various days. That is to say, the only factors to affect the outcome should be those of the experiment itself and not outside influences. One can examine other conditions that might affect the experimental outcome for many examples and different types of experimental endeavors. Our experiment was run on two consecutive days (replicated) as noted in the last two columns of Table 8.4 (Exp. 1 for results on day 1 and Exp. 2 for results on day 2). The runs were actually randomized and executed in a different order on the second day. The reason for the replication will be discussed in the next section.

5. **Define the responses to be determined and the statistical analyses to be performed.**
   There are, of course, many responses in many experimental considerations. Van der Heyden et al. (2006) discuss the responses determined in their ruggedness

test such as (1) the percent recoveries of main compound (MC), and perhaps a related compound (RC), (2) the resolution (Rs) of the critical peak pair, which is MC and RC, (3) the capacity factor (k) of MC, (4) the tailing or asymmetry factor (Asf) of MC and others. Table 8.4 shows the experimentally obtained design values for the generic response, %MC, in our situation for two experiments that were performed on each of the eight runs at the factor levels noted. Since our focus is on determining which of the factors may most affect the variance of the outcome or results (last two columns of Table 8.4), it is wise to run the experiment twice or replicate it. The reason for such is to obtain better estimates of the effects of the factors and get a more accurate estimate of the measurement variability. This guards somewhat against the possible measurement shift between the running of two designs. In general, replication reduces variability in experimental results, increasing their significance and the confidence level with which a researcher can draw conclusions about an experimental factor. Thus, it increases the reliability of the results. As an example, note the results in Table 8.4. Most of the replicated differences in %MC between Experiment 1 and Experiment 2 are about 0.2 or less. However, run numbers 3 and 5 have a difference of 0.4 or 0.5. This shift represents a twofold increase compared to the other runs. Although the numbers are still very close, in the case of seeking precision in experimentation a value from just one experiment, if no other experiment had been done, would perhaps be misleading. Replication alerts us to this possibility. A bigger discrepancy is noted in run number 4 with a difference of 0.9.

6. **Perform the experiment and determine the responses.**
We now suppose that the eight runs each from Experiments 1 and 2 have been done, and we derive the %MC responses in the last two columns of Table 8.4. This is our data. Thus, we are ready to move to the next step to begin to determine which of Factors A–G, if any, have a significant influence on the variation, that is, calculate the effects.

7. **Calculate the effects.**
Although usually done on a computer, the mathematics to calculate the effect of any one factor is fairly straightforward and we do so here to demonstrate the procedure.

One can see from Table 8.4 that the experiment is run twice (results in the last two columns), where each factor, A through G, is sampled at its higher level $(+)$ 4 times and at its lower level $(-)$ 4 times.

For example, in the first column for factor A there are 4 +'s and 4 −'s.

The factors $A, B, C, \ldots, G$ are assumed independent or orthogonal to each other.

The factor A effect in the first experiment is $(99.9 + 101.8 + 101.6 + 98.3)/4 - (96.8 + 100.5 + 100.2 + 98.7)/4 = 100.4 - 99.05 = 1.350$.

The factor A effect in the second experiment is $(99.5 + 101.7 + 101.5 + 98.1)/4 - (96.5 + 100.7 + 99.7 + 97.8)/4 = 100.2 - 98.675 = 1.525$.

So the average effect of factor A, Analyst, from the two experiments is $(1.350 + 1.525)/2 = 1.4375$.

The difference of the effect of factor A between the two experiments is

$$d = 1.350 - 1.525 = -0.175. \tag{8.1}$$

This same procedure is done for each of the other six factors, B, C, D, E, F, and G, to get their average effect over the two experiments and their difference. Table 8.5 is the complete table of results of the average effect for each factor within an experiment and the difference of effects between the two experiments. We can revisit the idea of replication here. Often one uses the term "offset." In other words, since we are running or replicating two experiments, one expects the results of one to be offset slightly (different) from the other. This may be due to some measurement error in the process of running the experiment. Since we are considering the same effects from the two sets of experiments, the statistically expected value between the effects of the two experiments is zero. This is seen in the last column of Table 8.5. Note that the expected value or average of d across the runs in Table 8.4 is close to zero. The actual value is -0.025 in the last column of Table 8.5. This is an indication that perhaps the replication was carried out appropriately.

8. **Derive the statistical and/or graphical analysis of the results.**
   Recall we want to know which of these factors most strongly influences the overall test, HPLC, measurement outcome. Thus, we need to compute a test statistic to help us determine if any one factor statistically influences the outcome. In this context, it is usually a t-statistic. To construct that statistic, we

TABLE 8.5   Ruggedness Analysis: Average Effect for the Two Levels for Each Experiment

Factor	Level		Experiment 1		Experiment 2		Difference (d)
			Average	Effect[a]	Average	Effect[a]	between Effects[b]
A	Analyst 2	+	100.4	1.35	100.2	1.525	−0.175
A	Analyst 1	−	99.05		98.675		
B	27	+	99.7	−0.05	99.275	−0.325	0.275
B	23	−	99.75		99.6		
C	Manufacturer 2	+	101.025	2.60	100.9	2.925	−0.325
C	Manufacturer 1	−	98.425		97.975		
D	40	+	100.15	0.85	99.20	0.475	0.375
D	30	−	99.3		100.9		
E	0.250	+	100.175	0.90	99.0	0.875	0.025
E	0.230	−	99.275		99.675		
F	7.3	+	99.825	0.20	99.575	0.275	−0.075
F	6.3	−	99.625		99.3		
G	Reagent 2	+	99.725	0.00	99.50	0.125	−0.125
G	Reagent 1	−	99.725		99.375		

[a]Effect = Difference between the two levels (+,−) within a factor.
[b]Experiment 1 effect minus Experiment 2 effect.

proceed as follows. Let us continue with factor A. We established that the difference in its effect between the two experiments is $d = -0.175$. We need to determine the overall standard error that is

$$S = \sqrt{\left[\sum d^2 / (N - 1)\right] (N/8)}, \qquad (8.2)$$

where $N =$ number of runs (in our case $N = 8$) and $\sum d^2 =$ sum of the squared difference of effects between the two experiments. In our case, this is the sum of the seven squared differences from each of the seven factors noted in the last column of Table 8.5. The value is 0.3744. Thus, for our experiment

$$S = \sqrt{[0.374/(8-1)](8/8)} = 0.2313. \qquad (8.3)$$

The desired t-statistic for testing each factor has the form

$$t = \frac{\text{Average effect of the factor}}{2S/\sqrt{2N}} \qquad (8.4)$$

on $N - 1$ degrees of freedom or

$$t = \frac{\text{Average effect of the factor}}{2\,(0.2313)/\sqrt{2 \times 8}}. \qquad (8.5)$$

Thus, for our case for factor A,

$$t = \frac{1.4375}{2\,(0.2313\,)/\sqrt{16}} = 12.429. \qquad (8.6)$$

The critical value for a t distribution on 7 (i.e., $N - 1 = 8 - 1 = 7$) degrees of freedom $= 2.365$

Thus, since 12.429 is obviously greater than 2.365, A is a significant factor or one that really may influence the outcome of the experiment. Some factors may have large t-statistics.

9. **Generate an appropriate report of the findings from the statistical analysis.**

See Table 8.6 for a summary of average effects and their t-statistics. Note that in examining this table the factors with a t-value greater than the absolute critical value of 2.365 are A, C, D, and E. All these t-statistics are greater than 2.365, which translates to $p <= 0.05$ and implies that A (analyst), C (column manufacturer), D (column temperature), and E (buffer concentration) somehow significantly influence the variability of the experimental outcome. One also begins to see the possibility of these factors being significant if one examines the effect columns of Table 8.5. Clearly, these four factors consistently have the larger effects across both experiments. This is not statistical evidence. However, it alerts one that these factor effects are certainly larger than the others. These factors must be better controlled for future experimentation, especially if the future goal is to reduce interlaboratory experimentation.

**TABLE 8.6   Ruggedness Experiment: Table of
Average Factor Effects and Their *t*-Statistic**

Factor	Average Effect	*t*-Statistic
A	1.4375	12.429
B	−0.1875	−1.622
C	2.7625	23.891
D	0.6625	5.729
E	0.8875	7.675
F	0.2375	2.054
G	0.0625	0.541

8.4   AN ALTERNATIVE PROCEDURE FOR RUGGEDNESS DETERMINATION

In this section, we are going to consider an alternative graphical approach to determining ruggedness (Torbeck 1996). Here we apply the method to a ruggedness design. As before, in some experimental situations we want to know if there is undue influence of factors on the outcome when we consider the combined effect of the factors at the same time. Table 8.7 is the result (effects) of a seven factor (A–G) eight-run experiment. Note this is the exact same setup as we had before, but with only one experiment.

Using the previous example in Section 8.3, recall that the factor A effect in the one experiment is $(+97.3 + 95.4 + 97.1 + 95.7)/4–(99.5 + 98.7 + 96.3 + 98.9)/4 = 96.375 − 98.350 = −1.975$.

This is just a matter of summing the "+" effects and subtracting the "−" effects and dividing by 4, which is the total of the number of plus effects and number of minus effects separately. Table 8.8 is the effect of all the factors from this one experiment.

TABLE 8.7   Plackett–Burman Design for a Single Experiment

Run	Ruggedness Factors							Results %MC
	A	B	C	D	E	F	G	Exp. 1
1	+	+	+	−	+	−	−	97.3
2	−	+	+	+	−	+	−	99.5
3	−	−	+	+	+	−	+	98.7
4	+	−	−	+	+	+	−	95.4
5	−	+	−	−	+	+	+	96.3
6	+	−	+	−	−	+	+	97.1
7	+	+	−	+	−	−	+	95.7
8	−	−	−	−	−	−	−	98.9

TABLE 8.8   Factor Effects from Table 8.7

Factor	Factor Effect	Absolute Value of Factor Effect
A	−1.98	1.98
B	−0.32	0.32
C	1.58	1.58
D	−0.08	0.08
E	−0.875	0.875
F	−0.575	0.575
G	−0.825	0.825

The last column of Table 8.8 is the absolute value of these effects. We'll explain that as we outline our procedure as follows. Much is the approach we took in Section 8.3, but repeat here for completeness.

1. Note that Table 8.7 is our design derived by the methodology we discussed earlier.
2. In Table 8.8, we have the factor effects in the center column and have merely rewritten their absolute value in the right column.
3. In Table 8.9, we order the absolute values in ascending order from the first row to the last row. They rank obviously from 1 to 7 as is seen in the third column of the table.
4. In the last column of Table 8.9 is the standardized rank computed from the ranks in the third column. These standardized ranks are easy to compute. Note that the mean of the ranks from one to seven is 4 with standard deviation $= 2.16$. Thus, the standardized value of 1 in the first row of the third column is computed as $(1-4)/2.16 = -1.39$. We do this for each of the other ranks from 2 to 7. The result is the last column of Table 8.9.
5. Next, construct a plot with the standardized ranks on the vertical axis and the absolute value of the effects on the horizontal axis as is seen in Figure 8.2. The statistical conclusion is that if the experimental method is rugged, the calculated factor effects will be normally distributed random noise and the normal probability plot will be a straight line.

The effects in Figure 8.2 appear to be modeled by a straight line. One may perhaps conclude that the method is rugged for the factors over the ranges tested. The rule is that if one or more points clearly do not lie on the line, one can conclude that the method is not rugged or robust for those factors. Those effects deviating from the linear fit of the line should be investigated for improvement or change of specification after they were considered to be of practical importance by an experienced investigator. In Figure 8.2, the coordinates of the point further from the line are (0.875, 0.46), which conforms to Factor E in Table 8.8. This may be a cause for concern. Here we may wish to examine the statistical significance of the effect of E in this experiment,

TABLE 8.9   Rank of Factor Effects from Table 8.8

Factor	Absolute Value of Ranked Factor Effect	Rank	Standardized Rank
D	0.08	1	−1.39
B	0.32	2	−0.93
F	0.575	3	−0.46
G	0.825	4	0
E	0.875	5	0.46
C	1.58	6	0.93
A	1.98	7	1.39

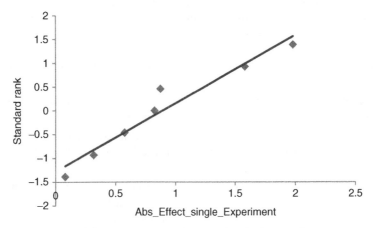

Figure 8.2   Normal Probability Plot for Single Experiment

which means deriving the *T*-test of the effect. The *T*-test for the single experiment effect of E is

$$T_{N-1} = \frac{\text{Effect(E)}}{2S/\sqrt{N}}, \tag{8.7}$$

where

$$\text{Effect(E)} = (97.3 + 98.7 + 95.4 + 96.3)/4 - (99.5 + 97.1 + 95.7 + 98.9)/4$$

$$= -0.875, \tag{8.8}$$

$N=8$ or the number of runs and after a bit of algebraic derivation S is actually the estimate of the pooled standard deviation of the (+) effects and (−) effects of the results in the last column of Table 8.7. Thus,

$$T_7 = \frac{-0.875}{2(1.582)/\sqrt{8}} = -0.7822. \tag{8.9}$$

The value of 0.7822 is certainly not greater than the critical t-value of 2.365 on 7 degrees of freedom. Thus, the method is rugged for all factors in this single experiment. The plot alerted us to check this factor as we did in equations (8.7) and (8.9).

8.5   RUGGEDNESS AND SYSTEM SUITABILITY TESTS

Vander Heyden et al. (1999) discuss the relationship between system suitability tests (SST) and robustness. An SST is an important part of many analytical methods. Its purpose is to determine the suitability and effectiveness of the operating system for certain expected response parameters for a particular method. That is to say, there may be expected limits on responses, such as %MC we discussed earlier, under certain conditions of the method. Thus, is the system suitable for the method we are performing, and how can one test that? The SST limits historically are established based on the experimental results obtained during the optimization and validation of a method and on the experience of the analyst. Obviously, from our earlier discussions, it is possible to evaluate the effects of the examined factors on different responses. Some parameters evaluated in an SST, such as resolution, peak tailing, and capacity factor, can also be considered as responses in a robustness test. In the way of a quick review, resolution is the quantitative measure of how well two elution peaks can be differentiated in chromatographic separation. Peak tailing is an asymmetry factor or chromatographic peak shape distortion. A peak is classified as tailing if its asymmetry is greater than 1.2, although peaks with asymmetry or skewness values as high as 1.5 are acceptable for many assays. The capacity factor is the migration rate of the analyte through the column. In setting up the initial suitability criteria in method development, the current FDA guidelines on "Validation of Chromatographic Methods" (1994) give the following proposed (recommended) acceptance limits as initial SST criteria: capacity factor > 2, resolution > 2 and tailing factor < = 2. These limits may change depending on the analyte or compound being investigated. Other recommendations include injection precision of %coefficient of variation (%CV) < 1% for number of replicates, $n >= 5$ and theoretical plate number > 2000.

The ICH (1994) guidelines state that "one consequence of the evaluation of robustness should be that a series of system suitability parameters (e.g., resolution tests) is established to ensure that the validity of the analytical procedure is maintained whenever used." Therefore, Vander Heyden et al. (1999) used the results of the worst-case situations (the scenario or factor levels that reflect minimum performance standards used to ensure that the chromatography is not adversely affected) to define the SST limits, for example, for resolution or capacity factor. Their recommendation is to determine the SST limits only when the method can be considered robust/rugged for its quantitative assay. In that case, it could be expected that in none of the points of the experimental domain, including those at which certain (system suitability) responses have their worst result, there would be a problem with the quantitative response.

To demonstrate the statistical procedure for the determination of suitability limits, we refer to the work of Vander Heyden et al. (1999) in which they investigated a ruggedness evaluation of a high-performance liquid chromatography (HPLC) method for identification and assay of ridogrel and its related compounds in ridogrel oral film-coated tablet simulations. We are going to follow their experimental methodology. They define the SST limit as the upper or lower limit from the one-sided 95% confidence interval around the worst-case mean. In other words, they determine which factor level combination (i.e., which run) from their design violates what is believed to be the preset SST limit. They then replicate this worse-case run about three times to derive a worst-case mean and then put a 95% confidence interval on that mean. If the SST limit is defined as being greater than a response value (e.g., resolution or capacity factor), then the lower limit of the 95% confidence interval for the worst-case mean is considered to be the SST limit. If the SST limit is defined as being less than a response value (e.g., tailing factor), then the upper limit of the 95% confidence interval for the worst-case mean is considered to be the SST limit. We will demonstrate this procedure as well as will introduce a second method to show how one can predict the worst-case mean with limits statistically from the experiment.

8.5.1   Determining the SST Limits from Replicated Experimentation

We proceed with the aforementioned HPLC experiment where the ridogrel is the MC. The capacity factor will be considered the response of interest. The original design was a 12-run Plackett–Burman design. There were eight factors in this design, which are given in Table 8.10.

Our design will follow that of Vander Heyden et al. (1999), but confined to the one response of capacity factor with an SST lower limit of 3.4 with a slightly different design and factor limits. The design levels of the 12-run 8-factor design are given in Table 8.11. We assume that the experiment is run and note the responses in the last column of the table. In the last column, the capacity factor ranges from 3.15 to 5.82. Note that two of the responses from experiment run 7 and 9 are both below the SST limit set earlier at 3.4.

TABLE 8.10   A List of Factors That Could Be Considered During Ruggedness Testing in the 12-Run+ HPLC Experiment

Factor	Label	Units	Level (−)	Level (+)
PH of the Buffer	A	–	6.3	6.9
Column Manufacturer	B	–	Manufacturer 1	Manufacturer 2
Column Temperature	C	°C	21	31
% Begin organic solvent[a]	D	%	23	25
% End organic solvent[a]	E	%	40	44
Flow of the mobile phase	F	ml/min	1.3	1.5
Detection wavelength	G	Nm	260	265
Concentration of Buffer	H	%m/v	0.220	0.270

[a]Percentage organic solvent in the mobile phase at the start and end of the gradient

TABLE 8.11   Plackett–Burman Design for the 8-Factor 12-Run Single Experiment

Run				Factors					Ruggedness Results: Capacity Exp. 1
	A	B	C	D	E	F	G	H	
1	+	−	−	−	−	+	+	−	3.810
2	−	+	+	−	+	−	+	−	5.820
3	−	+	−	−	+	+	−	+	5.245
4	−	+	−	+	+	+	+	−	3.790
5	+	+	+	−	−	+	+	+	5.075
6	+	−	+	−	+	−	−	+	3.995
7	+	−	+	+	+	−	−	+	3.150[a]
8	+	+	−	+	−	−	−	+	5.349
9	+	−	+	+	+	+	−	−	3.199[a]
10	−	+	+	+	−	−	+	−	4.682
11	−	−	−	+	−	+	+	+	3.378
12	−	−	−	−	−	−	−	−	4.941

[a]Note capacity < SST limit of 3.4. Worst-case scenario is observation 7.

Response run 7 at capacity factor = 3.150 would be considered the worst-case scenario in this context. The next step is to determine which of the factors in the experiment are significant. The effects can be computed with the same methodology we used earlier with the denominator of 6 for the + factors and − factors, that is, the effects of each factor (A–G) on the response, capacity, are calculated as

$$\text{Effect}_{(\text{Factor})} = \frac{\sum Y(+1)}{n} - \frac{\sum Y(-1)}{n}, \tag{8.10}$$

where $\sum Y(+1)$ and $\sum Y(-1)$ are the sums of the responses where the factor was at level $+1$ and at the level -1, respectively. The value, n, is the number of runs from the design, where the factor is at level $+1$ and -1. In our case, $n = 12/2 = 6$ since we have six runs for the lower level of the factor and six runs for the upper level of the factor. Recall from Table 8.11, we have a total of 12 runs for this experiment. Each factor effect and their statistical significance are given in Table 8.12. To determine the statistical significance of each factor, we can use the procedure of equation (8.7), which can easily be done by hand for each factor separately or we can create an analysis of variance (ANOVA) table as done by Vander Heyden et al. (1999). This is more computationally intensive and requires a computer program. The advantage is that all the factors are considered together and you reduce the risk of committing a type I error that we discussed in Chapter 2.

Stated simply, every time you conduct a t-test there is a chance that you will make a type I error, which is usually 5%. Therefore, by running two t-tests on the same data, you will have increased your chance of "making a mistake" to 10%. Performing three t-tests would increase the chance of a type I error to 15% and so on. These

TABLE 8.12　Twelve-Run, Eight-Factor Experiment

Factor	Label	Effect	P-Value
pH of the buffer	A	−0.547	0.5567
Column manufacturer	B	1.249	0.0205
Column temperature	C	−0.099	0.9953
% Begin organic solvent	D	−0.889	0.0413
% End organic solvent	E	−0.340	0.2409
Flow of the mobile phase	F	−0.573	0.3022
Detection wavelength	G	0.113	0.3324
Concentration of the buffer	H	−0.009	0.7937

Table of factor effects and their p-values.

are unacceptable errors. An ANOVA controls for these errors so that the type I error remains at 5% and you can be more confident that any significant result you find is not just by chance. The p-values in Table 8.12 are from the ANOVA. Using individual t-tests via equation (8.7) is a good first step to get some idea of which factors could be significant. Using an ANOVA would be the final word on which factors are in fact causing significant variation on the response outcome. Also, examining the table note, the two factors with the largest "Effect" are factor B or Column manufacturer and factor D or the percentage organic solvent in the mobile phase at the start of the gradient. Their effects are 1.249 and −0.889. These larger effects are also an indication of possibly significant factors. This is, of course, consistent with the p-values in the last column of the table of 0.0205 and 0.0413. Thus, these two significant factors will contribute most to the fact that in the worst-case scenario the SST limit of 3.4 in Table 8.11 is violated in observation 7. That is to say, this is the statistically significant factor level combination leading to the worst result for the response of the capacity factor. Note in Table 8.11 that these two factors of column and percent solvent are set at their lowest and highest levels, respectively. That is, the column is from manufacturer 1 and the percent solvent is at 25 (Table 8.10). As stated earlier, we wish to put a confidence interval on the worst-case mean. That is to say, we wish to compute the SST limit based on this experiment.

Since we have only one run in which the worst case is the lowest, that is, run 7 in Table 8.11, and the mean is the average of several runs, then one replicates the experiment a number of times for only those two factors, that is, the lowest column setting (manufacturer 1) and highest percent beginning percent solvent (25), which is much the same procedure as Vander Heyden et al. (1999). We have actually simulated the experiment five times for those two factor settings, which yielded the capacity factor results of 3.633, 3.174, 3.051, 3.491, and 3.229. Thus, the worst-case mean is $\overline{X}_{\text{WORST}_{\text{CASE}}} = 3.316$. If the SST limit is an upper limit, then we compute the one-sided lower confidence limit of the worst-case mean. This takes the form of

$$\overline{X}_{\text{WORST}_{\text{CASE}}} - t_{\alpha,\, n-1} \left(\frac{S}{\sqrt{n}} \right), \qquad (8.11)$$

where $t_{\alpha,n-1}$ is the critical t-value for significance level α on $n-1$ degrees of freedom and s is the standard deviation.

If the SST is a lower limit, then we compute the one-sided upper confidence limit of the worst-case mean. This takes the form of

$$\overline{X}_{\text{WORST}_{\text{CASE}}} + t_{\alpha,n-1} \left(\frac{s}{\sqrt{n}} \right). \tag{8.12}$$

Since the capacity factor is based on a value required in this setting to be above 3.4, then the new SST limit would be the lower one-sided limit of the worst-case mean. The one-sided critical t-value based on $n-1 = 5-1 = 4$ degrees of freedom is 2.132. The standard deviation is 0.239, thus the SST limit in this case is

$$\overline{X}_{\text{WORST}_{\text{CASE}}} - t_{\alpha,n-1} \left(\frac{s}{\sqrt{n}} \right) = 3.316 - 2.132 \left(\frac{0.239}{\sqrt{5}} \right) = 3.088. \tag{8.13}$$

Thus, this would lead to a system suitability limit of 3.09 for the capacity factor.

8.5.2 Determining the SST Limits from Statistical Prediction

We now discuss an alternative approach for determining the SST limits. We do this through statistical prediction from the 12-run experiment that we generated and the statistically significant factors that were noted. Instead of replicating the experiment at the factor settings that generated the worst-case scenario, that is, run 7 in Table 8.11, one regenerates the ANOVA for only those two factors, that is, the column setting (manufacturer 1) and percent beginning percent solvent, which were the significant factors of the original ANOVA. One then derives a predicted equation from the two factor ANOVA and inserts the factor-level settings (-1 for column and 25 for beginning percent solvent) to get a predicted worst-case scenario. This is much the same procedure as Vander Heyden et al. (1999). We demonstrate how this is done.

In general, the predicted equation for a response can take the form

$$Y = E_{F_1} F_1 + E_{F_2} F_2+, \ldots ,+ E_{F_k} F_k, \tag{8.14}$$

where E_{F_i} represents the effect of the factor considered for the worst-case experiment and F_i the level of this factor, $i = 1, \ldots ,k$. Note that this model does not contain a constant or take the form

$$Y = \beta_0 + E_{F_1} F_1 + E_{F_2} F_2+ , \ldots ,+ E_{F_k} F_k, \tag{8.15}$$

where β_0 is the constant term just as we encountered in the regression equations of Chapter 3. This is an important consideration as equation (8.14) may not be appropriate if the constant term is significant. In such a case, equation (8.15) may be more appropriate for prediction purposes. We will check all this out.

Recall from the original all factor ANOVA, in our case $k = 2$, we have only two significant factors. So how do we construct a predicted equation based on these two factors? First, we rerun the 12-run ANOVA with only these two factors, that is, column and percentage of organic solvent from the start (i.e., factors B and D in Table 8.11) to get a predicted equation. All the other factors are ignored since their p-values from Table 8.12 are greater than 0.05, and thus they have an insignificant contribution to the variation in the response. As an added statistical note here, one will often include factors with p-values less than 0.1 as they may come close to statistical significance and may be important from a knowledge-based perspective. Such is not the case in our situation as the two p-values take on the results 0.0215 and 0.0413. We, thus, proceed accordingly and run the models equations (8.14) and (8.15) with $k = 2$. The results are seen in Table 8.13.

We first run the analysis assuming equation (8.14) or no constant term. Recall that the response, Y, is the constancy factor. The worst-case scenario from run 7 in Table 8.11 and referring to factor levels in Table 8.10 is column, $F_1 = -1$ and percent organic solvent, $F_2 = 25$. Thus, from Table 8.13, we have $E_{F_1} = 0.624$ and $E_{F_2} = 0.181$. The predicted equation is, thus,

$$Y_{\text{Pred}} = E_{F_1} F_1 + E_{F_2} F_2$$
$$Y_{\text{Pred}} = 0.624 \ (-1) + 0.181(25) = 3.91. \tag{8.16}$$

Therefore, the predicted capacity factor from the worst case is 3.91. This is hardly a worst-case prediction as it is above the preset SST limit of 3.40. However, note the last row of Table 8.13. One can compute what we call a 95% prediction interval. Note that this interval is (1.875, 5.926). This is standard output from a regression analysis program and is very cumbersome to compute by hand. Thus, we do not attempt to compute it here. This is done in Chapter 6. The point is that based on this predicted equation of (8.14), the system suitability lower limit for the capacity factor from this experiment would be 1.875, which is certainly below the preset limit of 3.40.

Now let's examine the last two columns of Table 8.13. Note in the top row that testing the constant term not equal to zero leads to the p-value $= 0.002$. Thus, one

TABLE 8.13   Twelve-Run, Two-Factor Experiment

Factor	Label	(8.14) No β_0	p- Value	(8.15) Yes β_0	P- Value
		Model			
Constant term (β_0)		—	—	15.045	0.002
Column manufacturer	B	0.624 (E_{F_1})	0.027	0.624 (E_{F_1})	0.002
% Begin organic solvent	D	0.181 (E_{F_2})	0.001	−0.445 (E_{F_2})	0.014
Predicted constancy factor		3.91		3.29	
95% prediction limits		(1.875, 5.926)		(2.013, 4.589)	

Table of factor effects and their p-values.

would reject the absence of the constant term and the value of 15.045 may be reasonable. Therefore, equation (8.15) may be more appropriate than equation (8.14) for predicting the worst-case SST limit. Therefore, equation (8.15) combined with the parameter values given in Table 8.13 becomes

$$Y_{\text{Pred}} = \beta_0 + E_{F_1} \ F_1 + E_{F_2} \ F_2$$
$$Y_{\text{Pred}} = 15.045 + 0.624 \ (-1) - 0.445(25) = 3.296. \tag{8.17}$$

Thus, the predicted capacity factor from the worst case is 3.30. This is a worst-case prediction as it is below the preset SST limit of 3.40. However, note the last row of Table 8.13. One computes a 95% prediction interval as we did earlier. This interval is (2.013, 4.589). So based on this predicted equation of (8.15), the system suitability lower limit for the capacity factor from this experiment would be 2.013, which is certainly below the preset limit of 3.40. Also, note that the prediction in this case has greater precision than the case of equation (8.14) as the interval (2.013, 4.589) has narrower width than the interval (1.875, 5.926). Thus, one can see that it is important to not just go through the motions and do the computations of these predicted equations, but to check for their reasonableness and statistical validity as we did for checking the significance and appropriateness of the constant term in equation (8.15). Thus, putting together the information from the approach of equation (8.13) and lower prediction limit of equation (8.17), one sees a lower limit range of the capacity factor to be from 2.013 to 3.088.

REFERENCES

Becalski and Seaman with Health Canada (Becalski A and Seaman S. (2005) Journal of AOAC International. 88(1): 102–106.

Department of Transportation, Federal Highway Administration. (2012). Concrete Mixture Optimization Using Statistical Methods. Publication Number: FHWA-RD-03-060. http://www.fhwa.dot.gov/publications/research/infrastructure/pavements/03060/chapt2.cfm.

FDA Reviewer Guidelines: "Validation of Chromatographic Methods". November, 1994. http://www.fda.gov/downloads/Drugs/Guidances/UCM134409.pdf.

ICH Harmonised Tripartite Guideline prepared within the Third International Conference on Harmonisation of Technical Requirements for the Registration of Pharmaceuticals for Human Use (ICH), Text on Validation of Analytical Procedures. (1994). http:/www.ifpma.org/ich1.html.

ICH Harmonised Tripartite Guideline prepared within the Third International Conference on Harmonisation of Technical Requirements for the Registration of Pharmaceuticals for Human Use (ICH), Validation of Analytical Procedures: Methodology. (1996), 1–8. http:/www.ifpma.org/ich1.html.

Jain SP, Singh PP, Javeer S, and Amin PD. 2010. Use of Placket–Burman statistical design to study effect of formulation variables on the release of drug from hot melt sustained release extrudates. AAPS PharmSciTech. 2010 Jun; 11 (2): 936–944. doi: 10.1208/s12249-010-9444-6. Epub 2010 May 28.

John AE, Galea J, Francis SE, Holt CM and Finn A. (1998). Interleuken-8ˉm RNA expression in circulating leukocytes during Cardiopulmonary Bypass. Profusion13:409–417.

Plackett RL and Burman JP. (1946) The design of optimum multifactorial experiments, Biometrika33 (4): 305–325.

Torbeck, LD. (1996). Ruggedness and robustness with designed experiments. Pharmaceutical Technology. http://www.pharmtech.com/pharmtech/data/articlestandar/pharmtech/102011/710442/article.pdf.

United States Pharmacopeia 30. (2007) National Formulary 25. General Chapter 1225. Validation of compendial methods. United States Pharmacopeial Convention. Rockville, Maryland. USA.

Vander Heyden Y and Massart DL. (1996) Review of the Use of Robustness and Ruggedness in Analytical Chemistry; in A. Smilde, J. de Boer and M. Hendriks (Eds.) Robustness of analytical methods and pharmaceutical technological products, Elsevier, Amsterdam: 79–147.

Vander Heyden Y, Jimidar M, Hund E, Niemeijer N, Peeters R, Smeyers-Verbeke J, Massart DL and Hoogmartens J. (1999). Determination of system suitability limits with a robustness test. Journal of Chromatography A 845: 145–154.

Vander Heyden Y, Nijhuis A, Smeyers-Verbeke J, Vandeginste BG, Massart DL. (2001) Guidance for robustness/ruggedness tests in method validation. Journal of Pharmaceutical and Biomedical Analysis. 24(5–6): 723–753.

Vander Heyden Y, Nijhuisb A, Smeyers-Verbeke J, Vandeginsteb BGM, and Massart DL. (2006). Guidance for Robustness/Ruggedness Tests in Method Validation. http://www.vub.ac.be/fabi/tutorial/guideline.pdf

Yates F. 1935. Complex experiments. Journal of the Royal Statistical Society (Suppl); 2: 181–247.

INDEX

Introduction to Statistical Analysis of Laboratory Data, First Edition.
Alfred A. Bartolucci, Karan P. Singh, and Sejong Bae.
© 2016 John Wiley & Sons, Inc. Published 2016 by John Wiley & Sons, Inc.

Printed and bound by CPI Group (UK) Ltd, Croydon, CR0 4YY

16/04/2025

14658532-0001